OXFORD CLASSICAL MONOGRAPHS

Published under the supervision of a Committee of the Faculty of Literae Humaniores in the University of Oxford

OXFORD CLASSICAL MONOGRAPHS

The aim of the Oxford Classical Monographs series (which replaces the Oxford Classical and Philosophical Monographs) is to publish outstanding theses on Greek and Latin Literature, ancient history, and ancient philosophy examined by the faculty board of Literae Humaniores.

ROMAN NATURE

The Thought of Pliny the Elder

MARY BEAGON

CLARENDON PRESS · OXFORD
1992

Oxford University Press, Walton Street, Oxford OX2 6DP
Oxford New York Toronto
Delhi Bombay Calcutta Madras Karachi
Petaling Jaya Singapore Hong Kong Tokyo
Nairobi Dar es Salaam Cape Town
Melbourne Auckland
and associated companies in
Berlin Ibadan

Oxford is a trade mark of Oxford University Press

Published in the United States
by Oxford University Press, New York

British Library Cataloguing in Publication Data
Data available
ISBN 0–19–814726–0

Library of Congress Cataloging in Publication Data
Beagon, Mary.
Roman nature: the thought of Pliny the Elder / Mary Beagon.
— (Oxford classical monographs)
Originally presented as the author's thesis (Oxford University).
Includes bibliographical references and index.
1. Pliny, the Elder—Knowledge—Science. 2. Pliny, the Elder Naturalis historia. 3. Science—Europe—History. 4. Science, Ancient. 5. Philosophy of nature. 6. Man. I. Title. II. Series.
Q143.P64B43 1992 509.37—dc20
ISBN 0–19–814726–0

Set by Best-set Typesetter Ltd., Hong Kong
Printed in Great Britain by
Bookcraft (Bath) Ltd.
Midsomer Norton, Avon

PREFACE

THIS book is not intended to be a biography of Pliny. For information on his intriguing but ill-documented career and contacts the reader is referred to the series of articles by the late Sir Ronald Syme, in which a plausible life is painstakingly reconstructed. Nor is it concerned to cover in depth the various important specialist aspects of the *Historia Naturalis*. These too have been treated in other modern works: its chapters on art history, for instance, in a recent monograph by J. Isager; its scientific information on many occasions in the last few years, much of it at conferences and colloquia.

The aim of the present study is to view the *HN* as a whole; as testimony to the mode of thought, the outlook on life of the educated Roman citizen of the first century of our era. These rather general phrases are important. Too technical an examination of Pliny's intellectual heritage would be impossible, precisely because he is himself essentially a non-specialist. While he cannot in any sense be termed ordinary, he is not on the same level of intellectual innovation and artistic creativity as a Seneca or a Tacitus. Instead, he offers a valuable insight into the mind of a more typical member of the educated Roman élite. These were the absorbers rather than the innovators, for whom aspects of Greek learning and philosophy were worked into the fabric of their everyday lives, forming a backdrop of general principles and assumptions, rather than a platform for rigorous intellectual exploration. Thus, while I have found it possible to say a great deal in general terms about, for example, Pliny's philosophy and its effects on his ideas in the *HN*, to dwell too much upon possible sources and specific theories is ultimately unrewarding and for the most part irrelevant.

Instead, I have followed the lead given by Pliny himself, taking up the keynote of the *HN*, Nature, and in particular man's role in Nature, to build up his world view. Early chapters deal with his ideas on the cosmos, cosmic deity, and man's place in the cosmos. The second part deals with man's relation to specific areas of the natural world around him: the animal kingdom and the elements land and water. The final chapter looks at the use man makes of animals, minerals, but above all plants in prolonging and safeguarding his own life in Nature.

The *HN* is an impressively huge and unwieldy work. It is impossible to give it truly comprehensive examination in a book of ordinary length. The examples quoted to illustrate the various facets of Pliny's thought are by no means the only passages in which such views are expressed. Interested readers of Pliny's own writings will inevitably find equally apt and fascinating passages which I have mentioned fleetingly or not at all. To appreciate the range and versatility of Pliny it is in any case essential to work through the original. If the present survey encourages those previously daunted at such a prospect to overcome their hesitation, it will have achieved its purpose.

The idea of studying Pliny was originally suggested to me by my tutor at Somerville, Dr Miriam Griffin, who first encouraged me to undertake research. The successful completion of this book owes much to those who supervised the original thesis, Dr Barbara Levick, Professor Donald Russell, and especially Mr Nicholas Purcell, who offered general encouragement and devoted a great deal of time and effort to reading, criticizing, and suggesting numerous lines of inquiry which have invariably proved most rewarding. He and the late Dr Elizabeth Rawson also provided much helpful advice during the final revisions for publication, a process greatly aided by the many corrections and suggestions made by my D.Phil. examiners, Dr E. L. Bowie and Professor H. Hine, and, at a later stage, by Dr L. A. Holford-Strevens of OUP. Finally, my husband has been of great assistance in checking numerous drafts and revisions and has offered patient support throughout the composition process. My thanks are offered to all concerned.

September 1991 MARY BEAGON

CONTENTS

ABBREVIATIONS

THE following abbreviations occur in the notes and bibliography.

Reports

The four volumes of papers from the Pliny conferences at Como (1979) are referred to as follows:

Plinio e la natura	*Plinio e la natura: atti del ciclo di conferenze sugli aspetti naturalistici dell'opera pliniana: Atti della giornata di studi su Plinio e l'erboristeria* (Como, 1982)
Plinio il Vecchio	*Plinio il Vecchio sotto il profilo storico e letterario: atti del Convegno di Como, 5–7 ottobre 1979* (Como, 1982)
Tecnologia	*Tecnologia, economia e società nel mondo Romano: Atti del convegno di Como 27–9 settembre 1979* (Como, 1980)
La città antica	*La città antica come fatto di cultura* (Como, 1983)

Other works

'Pliny the Procurator'	R. Syme, 'Pliny the Procurator', *HSCP* 73 (1969), 201–36 (= *Roman Papers*, ii, ed. E. Badian (Oxford, 1979), 742–73)
'Carrière et amis'	Syme, 'Carrière et amis consulaires de Pline', *Helm.* 38 (1987), 223–31
'Consular Friends'	Syme, 'Consular Friends of the Elder Pliny', *Roman Papers*, vii, ed. A. R. Birley (Oxford, 1991), 496–511
Science in the Early Roman Empire	R. French and F. Greenaway (eds.), *Science in the Early Roman Empire: Pliny the Elder, His Sources and Influence* (London, 1986)
Helm. 37/38	papers given at the Nantes conference (May 1985), published as *Helmantica*, 37 (1986), 8–320, 38 (1987), 5–322; also as

	Pline l'Ancien, témoin de son temps (Salamanca and Nantes, 1987)
ANRW	*Aufstieg und Niedergang der römischen Welt*, ed. H. Temporini and W. Haase (Berlin and New York, 1972–)
Daremberg–Saglio	*Dictionnaire des antiquités grecques et romaines d'après le textes et les monuments*, ed. C. Daremberg and E. Saglio, 5 vols. (Paris, 1877–1919)
RE	*Real-Encyclopädie der classischen Altertumswissenschaft*, ed. A. Pauly, G. Wissowa, and W. Kroll (Stuttgart, 1893–1980)
SVF	H. von Arnim, *Stoicorum Veterum Fragmenta*, 4 vols. (Leipzig, 1903–24)

EDITIONS

For the most part, the Pliny text and translation in this book have been based on the easily obtainable edition of Rackham and Jones (Loeb). Any modifications have been made with reference to the Budé edition, or to the older texts of Detlefsen and Mayhoff.

Older editions

J. Sillig (Hamburg, 1850–5).
L. Jan (Teubner edn.; Leipzig, 1854–65).
D. Detlefsen (Berlin, 1866–73).
C. Mayhoff (Teubner edn.; Leipzig, 1899–1906).

Modern editions and commentaries

D. J. Campbell, *Naturalis Historiae liber secundus* (Aberdeen, 1936) [commentary].

H. Rackham and W. H. S. Jones (Loeb Classical Library edn.; Cambridge, Mass., and London, 1938–63) [ten compact volumes with English translation, though with less updated textual emendation than other modern editions].

J. Beaujeu *et al.* (Collection des Universités de France, Budé edn.; Paris, 1947– [most of the 37 books are now available with French translation and commentary; the latter vary somewhat in length and depth].

R. König and G. Winckler (Tusculum Bücherei edn.; Munich, 1973–) [text with German translation and commentary; vol. i conveniently includes the fragments of the lost works of Pliny].

F. Semi (Edizione Giardini, Pisa, 1977–) [text with Italian translation and brief commentary].

G. Conte, A. Barchiesi, and C. Ranucci, *Gaio Plinio Secondo, Storia Naturale*, 5 vols. (Turin, 1982–8).

Introduction

'He possessed an acute intellect, an amazing devotion to study, and an immense capacity for doing without sleep' (*acre ingenium*, *incredibile studium*, *summa vigilantia*). Pliny the Younger recalls in a letter (*Ep*. 3.5.8) how his uncle, besides long periods of study during the day, used also to work for at least half the night. He would then fit in a pre-dawn visit to the emperor Vespasian before beginning his other official duties. Even long periods of relaxation were only apparent; books were read and notes taken while bathing, eating, and resting. While travelling he always read, with a secretary constantly by his side to take notes. Walking, he admonished his nephew, was a waste of time. Compared with him, other men seemed enslaved to sleep and idleness. It comes as something of a relief to read that he was on occasion liable to drop off to sleep for short periods while working (*Ep*. 3.5.9). This picture of relentless activity complements Pliny the Elder's own words in the preface to the *Historia Naturalis*: 'life is being awake' (*vita vigilia est*, pref. 19). It is therefore somewhat ironic that we should remember Pliny for the manner of his death,[1] while commander of the Roman fleet at Misenum, in the eruption of Vesuvius that destroyed Pompeii and Herculaneum in AD 79. Yet it is understandable. His nephew's vivid description of those last days was written with the intention of firing the imagination of posterity. The historian Tacitus had requested an account of events in order

[1] The dramatic and detailed account given by Pliny the Younger (*Ep*. 6.16) has continued to provoke interest and speculation, even on the exact medical cause of his death. See R. M. Haywood, 'The Strange Death of Pliny the Elder', *Class. Week*. 46 (1952), 1–3. C. Zirkle, 'The Death of C. Plinius Secundus 23–79 AD', *Isis* 58 (1967), 553–9, and L. Bessone, 'Sulla morte di Plinio il Vecchio', *Rivista di studi classici*, 17 (1969), 166–79, argue that Pliny died of cardiac arrest rather than, as is generally concluded, asphyxiation, cf. Y. Grisé, 'L'illustre mort de Pline le Naturaliste', *REL* 58 (1981) 338–43, and M. D. Grmek, 'Les circonstances de la mort de Pline: Commentaire médical d'une lettre destinée aux historiens', *Helm*. 37 (1986), 25–43.

that he might leave an accurate description in his own writings. The dramatic tale of sensational catastrophe and heroism would enliven any history. Memorable in itself, it could be immortalized by suitable literary treatment.[2]

On the greater part of Pliny's life, however, we have lamentably little information. Attempts have been made to reconstruct his energetic career as soldier and statesman on the basis of the brief but tantalizing summary of Suetonius (*De Illustr.*) which refers to Pliny's *procurationes splendidissimae et continuae*. The number and location of these offices is not mentioned but passages in Pliny's sole surviving work, the *Historia Naturalis* (hereafter *HN*), may reflect personal experience of various provinces. The pioneering work by Münzer on this subject has been revised more recently by Sir Ronald Syme,[3] who has gone as far as is feasible in scrutinizing the evidence 'along with many things falling short of evidence'. On the basis of these discussions, a plausible career-sketch for Pliny may be reconstructed as follows.

He was born in AD 22 or 23 of an equestrian family at Comum in northern Italy. A number of society anecdotes in the *HN*, such as the spectacle of the empress Lollia Paulina encrusted with 40,000,000 sesterces' worth of pearls and emeralds at an ordinary betrothal banquet,[4] suggest that in his youth he may have moved in court circles at Rome during the principate of Caligula. His patron was Pomponius Secundus, whose biography he was to write (Pliny, *Ep*. 3. 5. 1). He had three spells of military service in Germany; with Corbulo against the Chauci in 47; in Upper Germany, where Pomponius was governor in 50, against the Chatti; and in Lower Germany again in 57–8, where he served with the future emperor Titus, the dedicatee of the *HN*, as prefect of a cavalry unit.[5] His experiences in this post

[2] See below, n. 37.

[3] R. Syme, 'Pliny the Procurator', *HSCP* 73 (1969), 201–36 (= *Roman Papers*, ii, ed. E. Badian (Oxford, 1979, 742–73), modifying F. Münzer, *Beiträge zur Quellenkritik der Naturgeschichte des Plinius* (Berlin, 1897); also 'Carrière et amis consulaires de Pline', *Helm*. 38 (1987), 223–31, and, most recently, with further modifications, 'Consular Friends of the Elder Pliny', *Roman Papers*, vii, ed. A. R. Birley (Oxford, 1991), 496–511.

[4] *HN* 9. 117. Perhaps 10. 12 and 14. 56 are also recalled from personal experience.

[5] An improved reading of the inscription on the *phalera* found at Vetera (*CIL* xiii. 10026)—'Plinio praef. eq.'—now confirms this (Syme, 'Consular Friends',

led to his first literary work, *On Throwing the Javelin from Horseback* (loc. cit.). It was while he was in Germany that he started the first of his larger-scale works, now lost, the *German Wars*.[6]

Pliny's activities from 59 onwards until Nero's death remain even more obscure. Unlike his remarks on other emperors in the *HN*, Pliny's comments on Nero are rarely neutral, never favourable. On the contrary, they are often unexpectedly violent in their condemnation. Pliny uses strong words in descriptions of even the most harmless frivolities.[7] It is possible that he was officially active, at least in the early years of Nero's reign, perhaps as procurator of Africa around 59; in which case, these remarks may have been prompted by feelings of guilt and self-justification. Pliny the Younger's description of the years between his uncle's youthful career in the courts and his death in 79—'throughout the intervening years, his time was much taken up with the offices he held and his friendship with the emperors'—may support this.[8] If he did retire from public life under Nero, he may have devoted himself to finishing his work on the German Wars. In the final years of Nero's reign, 'when the slavery of the times made it dangerous to write anything at all independent or inspired' (*Ep.* 3.5.3), he produced a six-volume work on the education of an orator and an eight-volume grammatical treatise.[9]

On the whole, it seems likely that Pliny rose to prominence under the powerful patronage of Vespasian and of Vespasian's

498), removing earlier doubts as to Pliny's status ('Pliny the Procurator', 207–8, '*Carrière et amis*', 227).

[6] On this lost work, see K. Sallmann, 'Der Traum des Historikers: Zu den *Bella Germaniae* des Plinius und zur julisch-claudischen Geschichtsschreibung', *ANRW* II. 32. 1, 578–601.

[7] H. Peter, *HRR* ii. 149; cf. *HN* 35. 51, 37. 50.

[8] *Ep.* 3. 5. M. T. Griffin, *Seneca: a Philosopher in Politics* (London, 1976), 438. See further below, pp. 17–18. It is, of course, possible that the Younger Pliny meant Claudius and Vespasian. He may even have been speaking loosely of Vespasian and his son Titus, for the latter was very closely associated in his father's government. The Elder Pliny himself refers to them as a father-and-son duo of rulers: '. . . Imperatores Caesares Vespasiani pater filiusque censores . . .' (*HN* 7. 162, cf. 2. 58; 3. 66). It is also conceivable that the nephew was being tactfully vague. For the possibility that Pliny was in Africa around 59 and not a decade later in the period of his other procuratorships, see Syme, 'Consular Friends', 502–3, but the question remains open.

[9] A. Della Casa, 'Plinio grammatico', in *Plinio il Vecchio*, 109–15.

adviser and general Licinius Mucianus, whose works Pliny often quotes.[10] Suetonius' phrase 'procurationes . . . continuas' suggests that the posts were held in rapid succession and thus could have fitted into the short period of ten years from 69 to 79. The 'eyewitness' passages from the *HN* point to three or four provinces: Narbonese Gaul, possibly Africa, Hispania Tarraconensis, and Gallia Belgica.[11] Only one of these is certain; we know from *Ep*. 3.5.17 that it was 'while he was procurator in Spain' that he met Larcius Licinus, *iuridicus* of Tarraconensis, who offered him 400,000 sesterces for his notebooks. Perhaps the most significant evidence in favour of any of these places comes when, in addition to a number of probable 'eyewitness' references (*vidi*, *vidimus*, etc.), Pliny also gives an exclamation of positive enthusiasm or a finely detailed descriptive passage. Thus, for instance, his praise of agricultural fertility in part of Africa (*HN* 18.188), or of a Spanish city distinguished with the laudatory epithet *urbs magnifica* in an otherwise dull gazetteer of bare names and population-figures.[12] Spain indeed receives praise from Pliny which is unsurpassed except by his descriptions of Italy herself. In *HN* 37.203, Spain is placed second only to Italy in richness of material resources. But Pliny goes further, adding that Spain has the advantage over Gaul in that even her deserts are productive: 'Spain is superior . . . by reason of her incentive to toil, her schooling of slaves, and the hardiness of body and eagerness of heart displayed by her inhabitants'. This passage above all suggests personal contact with Spain and her people.

There are also numerous references to the Spanish mines; Hispania Tarraconensis was a prime mining area.[13] The account of a mining method in *HN* 33.70–3 is painstakingly detailed. Pliny explains how arches are constructed to prevent the mountains from caving in on the mines. He explains the various methods of breaking through the barriers of flint and clay, and gives a vivid account of the final collapse of the rock (Ch. I.3).

[10] Syme, 'Pliny the Procurator', 208–9, notes that Pomponius Secundus was dead by 59 and there are no clues whether Pliny was taken up by another prominent noble under Nero. For further speculations on possible candidates see 'Consular Friends', 506–11.

[11] On the number and order of the posts, see Syme, 'Consular Friends' 502–5, and 'Pliny the Procurator', 211, on Münzer 103.

[12] Syme, 'Pliny the Procurator', 217.

[13] e.g. 3.3, 33.80, 34.144, 148, 164.

Such a passage suggests a real interest in the technical achievement here. Syme dates Pliny's procuratorship in Tarraconensis to 72–4, from the census figures for north-west Spain (3. 28), probably taken when Vibius Crispus, a man from Pliny's home area, was governor. If so, Pliny's visit to and interest in mines may have been semi-official. State control over the mines had been increased from the time of Augustus and procurators of mines appointed. It is possible that Vespasian took further measures in this direction.[14] Pliny could well have had official contact with the procurator of mines for the area or access through friends and connections in the province.[15]

Pliny's technical interests are of significance when we consider the nature of his *Natural History* and the social and literary background against which it was conceived. The first century of our era was a period in which the production of practical and technical manuals flourished. Their authors were predominantly members of the equestrian order, like Pliny himself, or 'new men' in the senate. Many of them came from the western provinces.[16] Such men lacked the prejudice of more aristocratic families against practical expertise. Finance had traditionally been a prerogative of men like Pliny. Their experience was of value in posts as financial procurators under the empire, and various technical accomplishments were an advantage in other important equestrian posts.[17] Pliny may have drawn upon the knowledge of certain such officials, obtaining information on grain yields and supplies from the Prefect of Rome's corn-supply (*praefectus annonae*) and samples of high-yielding grain from a procurator in Africa.[18] His own early military training had been

[14] Augustus: see T. A. Rickard, 'The Mining of the Romans in Spain', *JRS* 18 (1928), 129–43; J. Healy, *Mining and Metallurgy in the Greco-Roman World* (London, 1978), 112 ff. Vespasian: R. K. McElderry, 'Vespasian's Reconstruction of Spain; *JRS* 8 (1918), 94–101. Mining: see Healy, 'Pliny on Mineralogy and Metals', *Science in the Early Roman Empire*, 111–46; P. R. Lewis and G. D. B. Jones, 'Roman Gold-Mining in North-West Spain', *JRS* 60 (1970), 169–85; R. F. Jones and D. G. Bird, 'Roman Gold-Mining in North-West Spain', *JRS* 62 (1972), 59–74.

[15] Syme, 'Pliny the Procurator', 228–35.

[16] Ibid. 202 on equestrian interest and 219–25 on the writers' provenance.

[17] M. Frederiksen, 'Plinio il Vecchio e l'agricoltura in età imperiale romana: gli aspetti tecnici ed economici' in *Tecnologia*, 81–97.

[18] Such officials had to keep close watch on the quality of imports, cf Frederiksen (n. 17), who also quotes Sen. *Brev. Vit.* 18. 3, 19. 1. On grain samples, see *HN* 18. 94–5.

put to good use in his handbook on throwing the javelin from horseback. In fact, the various government posts open to equestrians both presupposed and encouraged such interests. Pliny's contemporary, the agriculturalist Columella, came of an equestrian family from Cádiz. He records in the preface to book 5 of his *De Re Rustica* how he was encouraged by his commander to continue his interest in agriculture and to take up land measurement as well (5. 1. 2–3). This latter skill bore fruit in the practical sphere when he was chosen to help found a colony near Tarentum.[19] Pliny himself seems, on the evidence of book 16 of the *HN*, to have had an interest in naval matters which predated his official post as admiral at Misenum. As we have already seen, he gained much information first hand through his tours of duty in the provinces. The posts themselves offered scope for furthering expertise and knowledge. Pliny reports (9. 62) how Optatus, his predecessor at Misenum, successfully restocked the sea between the mouth of the Tiber and the coast of Campania with wrasse from the Aegean.

Pliny cites equestrian writers of this period on numerous occasions. Besides Columella on farming,[20] we find, to name but a few, C. Julius Graecinus of Narbonensis, the father of Agricola (indexes to books 15–18, on agriculture); Turranius Gracilis of Gades, an expert on cereal (e.g. index to book 9, 18. 75); Cornelius Valerianus (10. 5, 14. 11); the geographer Pomponius Mela (indexes, especially to books 3–7); and Papirius Fabianus (e.g. 9. 25, 15. 4).[21] Seneca, himself of equestrian origins and the first of his family to enter the senate, is mentioned respectfully as *minime mirator inanium* (14. 51). However, sources such as these, identifiable by name, do not tell the whole story. It is clear from certain references in the *HN* that these writers' active curiosity on such matters as architecture, agriculture, geography, and natural sciences was shared by their class as a whole. Pliny cites 'distinguished members of the order of *equites*' as authorities for the statement that 'a sea-creature in every part of its body resembling a man has been seen in the Gulf of Cádiz . . .' (9. 10). In *HN* 5. 11 ff. he describes the quelling of rebellion in Mauretania

[19] *CIL* x. 235 = *ILS* 2923.
[20] Ch. V. 1. B. ii.
[21] See Syme, 'Pliny the Procurator', 219–25, on the origins and interests of these and other Plinian sources.

at the beginning of Claudius' principate. He testifies to the excitement not only of the senatorial commanders on that expedition who boasted of penetrating to the Atlas mountains but also of the *equites* who made similar claims when they subsequently governed the country. Indeed, these officials tend to give a false impression of the country's accessibility, 'since persons of high position, although not inclined to search for the truth, are ashamed of ignorance and consequently are not reluctant to tell falsehoods'. Pliny gives an impression of procurators vying with one another to render sensational descriptions of the latest explorations and discoveries because such accounts are now expected of them. In a description of the towns of Mauretania preceding this passage, Pliny ridicules an equestrian writer of the previous generation, Cornelius Nepos. He has swallowed 'with avid credulity' many tall stories and misleading statistics about the town of Lixus and its river (5. 1–4). If, at some stage in his career, Pliny did hold a procuratorship in Africa, he may well have considered himself something of an expert on that part of the world. The rivalry certainly sounds as if it has a personal edge in Pliny's sour conclusion: 'For my part, I am less surprised that some things are outside the knowledge of the equestrian order (some of whom indeed nowadays actually get into the senate) than that anything should be unknown to luxury' (5. 12).[22]

For the most part, it is men of the equestrian order who find that their practical writings complement rather than conflict with their official duties. Pliny, it is true, does tell his dedicatee Titus in the preface to the *HN* that *officium* takes first place with him. His writings are the product of whatever *otium* is left to him, and are therefore often a nocturnal pursuit (pref. 18). Yet perhaps he was bound to say this to the son of the emperor who employed him. For Pliny's class, *otium* and *officium* combine far more easily than they did for the conscientious senator of the Republic. Under the Principate, equestrians did not on the whole incur the imperial suspicions of rivalry and treason which dogged the highest order. For the senator, withdrawal into *otium* could be as risky as over-independent political activity.

The material collected by these writers commanded an en-

[22] Ch. II. 4.

thusiastic audience. The sum Larcius Licinus offered Pliny for the notebooks, as yet incomplete, in which he was collecting the material for the *HN* (*Ep.* 3.5), was the equivalent of the minimum equestrian census qualification. Those who reached the highest offices of state might continue to nurture such interests. Seneca was interested in agriculture,[23] and Licinius Mucianus wrote a book on curiosities drawing on his experiences as legionary legate and provincial governor.[24] The subject-matter of this latter work reflects another facet of the equestrian literary output; one which at first sight seems a rather unlikely choice for writers whose interests and occupations were so frequently practical and technical. Interest in natural and man-made oddities can be traced back to Homer's *Odyssey*. More recently, a tradition of writings specifically on *mirabilia* had developed in the Hellenistic age. Many of the authors known to have written such works are listed in the indexes of the *HN*. Callimachus and his pupil Philostephanus of Cyrene are cited in the indexes of a number of books. From the Roman era, Isigonus of Nicaea is listed, and the omnivorous scholar Alexander Polyhistor receives numerous acknowledgements; some at least may be for his *Collection of Wonders*. These and others are cited particularly frequently for book 7, which includes a large selection of human *mirabilia*. Most interesting of all are those passages in the main body of the *HN* where Pliny actually names particular authors as sources for specific stories. Thus Alexander is cited on such diverse natural oddities as an ageing process in particular types of pearls (9.115); an otherwise unknown species of tree he claims was used to build the *Argo* (13.119); and the diet of beech-mast eaten by the inhabitants of Chios during a siege (16.16–17). Isigonus and Nymphodorus provide stories of African tribes who practise sorcery and possess the Evil Eye (7.16), while Callimachus records the curative powers of a Phrygian river (31.9). The *HN* is in fact a valuable source for the study of such writings. In addition to direct citations, a large number of stories unattributed by Pliny can be plausibly assigned to these and other authors. In the case of one or two such writers, what little we know about

[23] M. Griffin (n. 8), 289–91.
[24] See 12.9 (Ch. II. 5), 13.88 and further Syme, 'Pliny the Procurator', 203.

them is due almost exclusively to conjecture based on passages of the *HN*.[25]

Such writings were not exclusively Greek. A number of Roman writers had also produced studies which are quoted extensively by Pliny. It is through him that we hear of the work on curiosities by Mucianus, and of the existence of other studies with similar emphases. For instance, 'the volume by Aristander teems with portents of this nature [connected with trees] in Greece, as do the *commentarii* of C. Epidius in our own country, including ones of trees that talked' (17. 243). Pliny's quotations of Turranius Gracilis (9. 11), Pompeius Trogus, Cornelius Valerianus, and other writers of his own class and epoch often reflect their interest in *mirabilia*. Trebius Niger included in his work on marine life a sensational description of a giant octopus in Baetica which would come inland at night and raid fish-ponds, scaling the fences erected to keep it out. It was eventually killed and its remains, presumably pickled, 'kept as a curiosity' (*miraculum*) by the governor (9. 92–3). Valerianus reported the sighting of a phoenix in Egypt in 36 (10. 5), while Turranius Gracilis attests a 'sea-monster' cast up on the shore at Cádiz (9. 11). The apparent incongruity with the technical, practical aspect so often encountered in such writers' work is perhaps to be explained by Pliny's comment in *HN* 11. 6: 'The more I observe nature, the less prone I am to consider any statement about her to be impossible'. The feeling that truth was stranger than fiction may well have encouraged an interest in *mirabilia* in the first century, in spite of the steady expansion of geographical knowledge, often

[25] The paradoxographical tradition continued into the second century. See A. Westermann, *Paradoxographi: Scriptores Rerum Mirabilium Graeci* (London, 1839); O. Keller, *Rerum Naturalium Scriptores Graeci Minores* (Leipzig, 1877); A. Giannini, *Paradoxographorum Graecorum Reliquiae* (Milan, 1965). Pliny's Index references include: Callimachus (bks. 4, 7, 9, 10, 11, 22); Isigonus (bks. 7, 12, 13); Alexander Polyhistor (bks. 2, 3, 4, 5, 6, 7, 9, 12, 13, 16, 36, 37). Direct citations; see (Giannini) Callimachus fr. 46, Nymphodorus fr. (8), Agatharchides fr. (2), Isigonus fr. 15, Alexander fr. (5), (6), (7). Parallels for fragments of a number of such authors may be found in Pliny; see Giannini on e.g. Antigonus. Authors who appear to be largely re-created from Pliny include Aristander (Giannini no. 23, pp. 369 ff.), [Orpheus] (no. 27, pp. 384 ff.) and Bolus (no. 30, pp. 377 ff.). Cf. also Ch. IV. 1. i. For Pliny's interest, see now J. Isager, *Pliny on Art and Society: The Elder Pliny's Chapters on the History of Art* (London, 1991), 45–7, but in particular his discussion of man-made wonders in Pliny, especially the buildings of Rome, 195–205.

through military penetration.[26] Pliny's comments on the new inroads into Africa reflect the interest in the nature and contents of a territory which was among the least familiar to the ancients (cf. 32. 143). The strangeness of what had been discovered heightened expectations of what was to come. Pliny famously recalls in *HN* 8. 42 the common saying of Greece that Africa is always producing some novelty (*vulgare Graeciae dictum semper aliquid novi Africam adferre*).[27] Fantastic stories about such lands may have originally developed because so little was actually known. Ironically, exploration may have enhanced rather than abolished such rumours. The pioneering knights mentioned by Pliny clearly felt they would lose face if they did not return with some facts and figures to satisfy an eager and curious audience at home. Accordingly, they would invent rather than admit ignorance. No doubt they were also predisposed to confirm or even better the predictions of marvels, rather than report objectively, especially if the truth was not as sensational.

Thus it was that interest in marvels flourished in, and was encouraged by, an atmosphere of optimism and expansion in the first century. In the cultural and literary sphere, however, the collection of marvels 'for their own sake' has been seen as a symptom of decline.[28] Certainly, interest in oddities could have an unhealthy aspect, as described by Plutarch in his treatise *On Curiosity*. He criticizes the sensation- and horror-seeking curiosity which leads some people to despise more normal interest in the heavens or the flowering habits of plants because they have nothing evil in them. Even the most sensational and bloodthirsty history is rejected because they 'prefer their calamities hot and fresh' (517 F). Instead of gazing at statues and paintings, they frequent the freak-shows to gape at hideous deformities. Not only will such sights eventually induce satiety and nausea, but they are a waste of the intellect and amount to indiscriminate destruction of the senses rather than intellectual discipline (521 B–D). Plutarch might well have condemned treatises devoted to wonders as being substandard literature.

[26] Ch. V. 2. c. and n. 48.

[27] Arist. *HA* 8. 28, 606^{b}20. See E. Bianchi, 'Teratologia e geografia', *Acme*, 34 (1981), 227–49.

[28] e.g. G. Williams, *Change and Decline: Roman Literature in the Early Empire* (Berkeley, Calif., 1978), esp. 190–2.

Yet it must be remembered that for many authors of Pliny's era such books seem to have been only part of their output, and that for others curiosities were only a subsidiary topic in a particular work. Pliny records many wonders and oddities of Nature, from monstrous births and deformed humans to mythical animals like the phoenix.[29] But the voyeurs described by Plutarch would have had a tedious job trying to pick out such references from the thirty-seven books of the *HN*. It is also clear from his own comments that, by ancient standards, Pliny was quite sceptical of many marvellous stories.[30] He compares well, for instance, with Pomponius Mela, and his reputation for credulity is due largely to later writers like Solinus and after him the authors of medieval bestiaries who picked out the Plinian stories but omitted the author's reservations.[31]

'My subject is nature, that is, life' (*HN*, pref. 13). It is, above all, the scope of Pliny's work which differentiates it from many other technical treatises which confined themselves to a single topic such as agriculture, medicine, or architecture, and links it with a classical tradition of encyclopaedic literature. The Greeks, it is true, had never compiled an encyclopaedia, and their attitude to learning was, in general, of a less practical nature, with more emphasis on specialized studies and discussions of partic-

[29] For Pliny, many true facts of nature are amazing. Many words indicative of surprise and wonder occur throughout the *HN*. The following are examples of what we should term freaks, oddities, or fantasies: *HN* 7 *passim*, esp. 9–21; 22–32 on the wonders of India and Ethiopia; 34–6 on monstrous births and hermaphrodites; 73–6 on dwarfs and giants; 8. 37, 80, 153; 9. 2; 11. 272; 16. 132; 17. 241, 244; 18. 166.

[30] The phoenix (10. 3) is 'possibly fabulous'. Stories of werewolves are just fantasy (8. 80). The scientific limitations of the era together with his idea that 'truth is stranger than fiction' make it very difficult for Pliny to decide what to accept and what to reject. Very often he is non-committal. Occasionally he decides to reject ideas which later scientific advances have proved to be true. He is, for example, sceptical of Trogus' assertion that palm-trees can be propagated from the leaves (17. 58). Cf. Herodotus 3. 115. 1; Strabo 7. 2. 1.

[31] Contrast, for instance, 10. 136 with Solinus 30. 29: Pliny remarks of the 'pegasus' bird, the griffin, and the 'bearded' bird *fabulosos reor*; Solinus mentions these creatures but without the disclaimer. See also G. Steiner, 'The Scepticism of the Elder Pliny', *Class. Week.* 48 (1955), 137–43 and L. A. Holford-Strevens, *Aulus Gellius* (London, 1988) for Gellius' part in misrepresenting Pliny. Pliny can also be misjudged as a result of misreading and misunderstanding on the part of the epitomators and later users of the *HN*. These errors can lead to a garbled misrepresentation of the original text: see *C. Iulii Solini Collectanea Rerum Memorabilium*, ed. Th. Mommsen (Berlin, 1895), esp. pp. viii–ix.

ular problematic questions. There was, however, a practical note in the claim of the Sophists to teach a man all the skills (*technai*, *artes*) that he might need in everyday life. Aristotle later called these *eleutheriai epistemai*, the basic knowledge for a properly educated free man. It was to this idea, of a non-specialized but wide-ranging knowledge which, as it were, makes its pupils properly educated and decent citizens, that Pliny alludes in his Preface: 'Before all else, we should touch upon what the Greeks called *enkýklios paideia*' (pref. 14). The Roman writer Quintilian was later to use the same phrase to denote a general education preliminary to more specialized studies.[32] The literary development of this idea, in the collecting of the various components of *enkyklios paideia* into a single volume or encyclopaedia, had, in fact, been a Roman achievement. A command of both words and actions, combined in and bonded together by a practicality of outlook, was regarded as characteristic of the Roman approach to life. In Roman eyes, the Greeks often overtheorized: 'What could be more presumptuous or a greater waste of words?', asked Cicero, describing how the Greek philosopher Phormio lectured the seasoned campaigner Hannibal on the military *ars*. Cicero's ideal orator was the cultivated man of action. His rhetorical studies should be supported by 'complete understanding of everything relevant to man's customs, to life, to virtue, and to the state'.[33] This attitude to knowledge was epitomized in the writings of Cato the Elder.[34] Philosophy is bound up with ancestral precepts and practical life, rather than with philosophical speculation. Knowledge is not pursued for itself, but follows the courses of activity and life. Life, according to Cato, was like iron, which rusted if inactive,[35] a thought which may be compared to Pliny's own saying, as recorded in the *HN*'s preface: 'Life is being awake' (pref. 19).

[32] Plato, *Hp. Mi.* 368 b, on Hippias of Elis; Arist. *Pol.* 1337^{b}15. See e.g. H. I. Marrou, *Saint Augustin et la fin de la culture antique* (Paris, 1938), 211–35 (cf. id., *Histoire de l'éducation dans l'antiquité*[6] (Paris, 1965, 266, 568 n.)). On the history of the phrase *enkyklios paideia* see H. J. Mette, 'Enkyklios Paideia', Gymnasium, 67 (1960), 300–7.

[33] *De Orat.* 2.75, 3.72.

[34] P. Grimal, 'Encyclopédies antiques', *Cah. d'hist. mond.* 9 (1965), 459–82, takes Cato to be the first exponent of the Roman encyclopaedic tradition. A. E. Astin, (*Cato the Censor* (Oxford, 1978), App. 8, pp. 332–40), while acknowledging the emphasis in the fragments upon practical knowledge, doubts whether they added up to anything as extensive and systematic as an encyclopaedia.

[35] Fr. 3 Jordan ap. Gell. *NA* 11.2.6.

Both these aspects, a Roman practicality and a comprehensiveness of scope, are prominent features of Pliny's *HN*. Indeed, they are developed to a point which actually sets the *HN* apart from apparently related works. Two encyclopaedias which preceded the *HN*, Varro's *Disciplinae* and Celsus' *Artes*, had their material arranged in books according to the various *artes* covered. By calling his book *Historia Naturalis*, 'inquiries into the natural world', Pliny gives it a single unifying theme. His readers will be educated for everyday life, by the examination of all aspects of the world around them: *natura, hoc est vita, narratur* (pref. 13, cf. p. 55). For Pliny, useful knowledge is not to be artificially divided and confined within the bounds of certain *artes*. In this respect it is misleading to see the *HN* in the context of encyclopaedic literature and its development from the subject divisions of Varro's and Celsus' works into the fixed structures of the medieval system of the seven liberal arts, the *trivium* and *quadrivium*. A more telling comparison might be made with one of the earliest exponents of the 'arts of living', the sophist Hippias of Elis. He is held up by Cicero for the edification of the ideal Roman orator: 'He claimed that there was not a single fact included in any subject with which he was not acquainted. He had not only acquired the *artes* which form the basis of the liberal education of a gentleman—mathematics, music, knowledge of literature and poetry, and the doctrines of natural science, ethics, and political science—but he had made with his own hand the ring he had on, the cloak he was dressed in, and the boots he was wearing.'[36] The Sophists had claimed to teach everything necessary for success in life. The whole subject and purpose of Pliny's work is 'life'. The structure of his inquiries is dictated by that of the natural world as viewed by man, starting with the cosmos as a whole, in book 2, and progressing through all its subdivisions, animal, vegetable, and finally mineral. It is within this unique structure that theory and scholarship are united with the practical arts of agriculture and seafaring, of government and warfare. Philosophy is united with practical philanthropy.

Inseparable from the encyclopaedic tradition is the idea of the educated man, who is neither an intellectual expert nor a specialist. In many respects the author of the *HN*, a well-

[36] *De Orat*. 3. 127.

educated and practical Roman of the first century, can himself be seen as an outstanding example of 'encyclopaedic man'. Such qualities encouraged Pliny the Younger to commit to history an image of his uncle as the ideal Roman, *vir bonus*.[37] His well-known account of Pliny's death includes the statement that the fateful expedition started as a scientific inquiry and ended as a humanitarian exercise (*Ep*. 6. 16. 9). There is, of course, an element of *pietas* in the younger Pliny's account, a desire to depict his uncle as the perfection of a 'type'. But a similar ideal is to be found in his uncle's self-presentation in his *HN*. Besides the whole tenor of practical learning in the *HN* and the unstinted praise of all things Roman, his praise and admiration for Cato the Elder are particularly significant in this context (*HN* 29. 13). Cicero, pre-eminent in both politics and literature, is also singled out for eulogy (7. 116–17). Pliny, who himself had practised at the bar in youth (*Ep*. 3. 5. 8), besides his military, political, and literary activities, clearly felt an affinity with the quintessentially Roman talent of his great predecessor.[38]

In so far as he can be regarded as a type, Pliny provides the student of cultural history with an invaluable balance to more overtly intellectual writers of this period, such as Seneca. The gifted and original exponents of new ideas are not typical of the mass of those who must have been influenced by such ideas. In order to widen our historical perspective, we must look beyond isolated study of the mind of the philosopher to the effect his theories had on the attitudes of contemporary society at large. Here the views of a more 'ordinary' subject, well educated but not esoterically intellectual, are of the greatest value. Pliny, with his encyclopaedic work, is an ideal example. The *HN* must, of course, also be viewed from a wider perspective than that of literary genre. It may, as we have seen, be compared with writings produced by other members of Pliny's class at about this time. It

[37] See Grisé (n. 1). K. Sallmann, '"Quo verius tradere posteris possis" (Plin. epist. 6, 16)', *WJA*, NF 5 (1979), 209–18, suggests that Pliny the Younger was not satisfied with accounts which portrayed his uncle as merely curious. In spite of problems with meteorological details and the exactitude of his uncle's medical symptoms, the letter, written just 30 years later, can hardly have been as fictitious as R. Martin, 'La mort étrange de Pline l'Ancien, ou l'art de la déformation historique chez Pline le Jeune', *VL* 73 (1979), 13–21, suggests.

[38] M. A. Burns, 'Pliny's Ideal Roman', *CJ* 59 (1964), 233–8, argues that M. Agrippa is treated very favourably in the *HN* and that, as soldier, statesman, and scholar, he too may have had a particular appeal for Pliny.

is also very much a product of its age, literally an epitome of the culture of the first century. It covers the subject-matter of the usual education of the ruling classes and is often couched in the rhetorical prose which is itself a prime feature of this education.[39] It can, moreover, be viewed as the end-product of developments in the last two centuries BC which aimed to bring Greek philosophy to a wider audience. Various literary treatises began to bring together the earlier Greek philosophical ideas, and to present them in a form comprehensible to this audience. In the West, Cicero's philosophical works and the wide-ranging and learned writings of Varro were particularly important in this respect. In the East, Philo of Alexandria approached Greek philosophy from a Jewish standpoint.

The effect of the spread of intellectual theories among generally educated people may be seen in the work of a writer like Pliny. While he writes in a spirit of inquiry and volunteers ideas and facts as being entirely new (e.g. 2.71, 'a theory recorded by nobody before myself'),[40] detailed philosophical argument for its own sake is not his purpose. Instead, his acquaintance with philosophical ideas moulds his thought in a general way and colours his attitude to everyday life. Unlike his near-contemporary, Seneca, he is not claiming to be a philosopher.

One effect of the popularization of philosophy was a tendency to eclecticism. In a non-specialist like Pliny, this is liable to be more complex because his grasp of the original ideas is less exact and vaguer. With regard to cosmology, he was certainly influenced, at least through an intermediary source, by Plato's *Timaeus*,[41] as were many writers both before and after his time.

[39] See e.g. D. A. Russell, 'The Arts of Prose: The Early Empire', in J. Boardman, J. Griffin, and O. Murray (eds.), *The Oxford History of the Classical World* (Oxford, 1986), 653–76. On Pliny, see A. Önnerfors, *Pliniana: In Plinii Maioris Naturalem Historiam Studia Grammatica, Semantica, Critica* (Uppsala, 1956) and D. J. Campbell's edn. of *HN* 2 (Aberdeen, 1936).

[40] See Ch. II. 1.

[41] A. S. Pease, 'Caeli ennarrant', *Harv. Theol. Rev.* 34 (1941), 163–200, stresses its general influence, perhaps aided by Posidonius' commentary (179). On its importance to Philo, see Colson's introduction to his Loeb edn. of *De Aeternitate Mundi* (Philo of Alexandria ix, Cambridge, Mass., 1941), 175. It was also of importance to the whole of the Hexaemeral tradition, from Philo's *De Op. Mundi* to the most important of the Christian *Hexaemera*, that of Basil, and thence to later writers such as Philoponus and Augustine (see F. E. Robbins, *The*

He also appears to have known the *De Natura Deorum* of Cicero well, and used it in other contexts.[42] With its dialogue form, this work in fact epitomizes the educational function of these later treatises,[43] setting out the main philosophical views on the cosmos and the divine for comparative and argumentative purposes.

Because of this background, detailed analysis of Pliny's philosophical knowledge is of limited value. An example will serve to show its generalized nature. Pliny says at the beginning of book 2 that the world is sacred, eternal, and immeasurable, at once a work of nature and nature herself, and is appropriately believed to be a deity. Aristotle held that the world was both uncreated and indestructible, Plato that it was indestructible but created. Indestructibility was not the common view among the Stoics, and those who held it were forced to abandon the Stoic doctrine of the periodic conflagration.[44] Pliny, however, also mentions *ekpyrosis* in a later book (7. 73). While such a fleeting reference need not be very significant, it is interesting to compare Philo of Alexandria's remarks in his *De Aeternitate Mundi* (7–11). According to the Stoics, he says, the world may be called from one point of view eternal, from another, perishable. In one sense, it is perishable in that it is reconstructed after a conflagration, in another, it is eternal because it goes through constant rebirths. Philo had begun his discussion by remarking on the problems caused by ambiguous terminology, in particular the many meanings of both *kosmos* (world) and *phthora* (destruction), *De Aet. Mundi* 3–5. It is hardly fair to blame Pliny too much for careless error or inconsistency. What we see is an imprecise absorption of such subtleties by a layman.

The second-hand nature of the philosophical components must

Hexaemeral Literature (Chicago, 1912), Ch. 1). In Pliny, there are undertones of the *Timaeus* overseer-god in his description of the sun as *rector* of the universe (2. 12), and in *HN* 2. 5: *mundus* is a sphere which 'sibi ipsa toleranda est, seque includit et continet', cf. *Timaeus* 34 B, 'a single spherical universe, alone, but because of its excellence, needing no company other than itself'.

[42] See A. S. Pease in his edn. (Cambridge, Mass., 1955), i. 53.

[43] The *De Aeternitate Mundi* of Philo is a good example, as is also the pseudo-Aristotelian *De Mundo*, probably written between 50 BC and AD 100. See E. S. Forster in vol. vii of the Oxford translation of Aristotle (1914), and W. Capelle, 'Die Schrift von der Welt', *Neue Jahrbücher*, 15 (1905), 632 (a possible basis in works by Posidonius).

[44] Philo, *De Aet. Mundi*, 7–11, 76 (*ekpyrosis*).

not obscure the fact that the picture Pliny builds up with them is uniquely his own. The originality of his integration of the traditional constituents of *enkyklios paideia* into a unity based on the theme of *Natura* has already been mentioned. Morever, the reader of the *HN* as a whole will be made aware of the individuality of its author. If he is, in many respects, a model of the educated but practical Roman statesman-scholar, he also possesses emphases in outlook peculiar to himself. On the subject of that epitome of Roman virtues, Cato the Elder, Pliny is by no means blindly uncritical. 'Cato . . . is deemed to have exemplified the three supreme human achievements, excelling alike as orator, as general, and as senator; all of which . . . seem to me to have been achieved with greater brilliance in the case of Scipio Aemilianus, and that, moreover, without the very wide unpopularity which handicapped Cato' (7. 100). Cato, he continues, took part in forty-four law-suits, and was sued (unsuccessfully) more often than anyone else. Pliny, it seems, personally values tact, diplomacy, and even compromise in public life. Cicero's opinion of Cato's great-grandson, the younger Cato, had, on occasion, been similar,[45] and, more recently, such qualities had become advisable in imperial politics.

As a loyal Flavian supporter, Pliny's hostility towards the emperor Nero (see above, p. 3) might have been no more than a political statement of solidarity with the new dynasty. Yet it is unlikely that effective flattery was his aim in the scattered references of so large a work as the *HN*. If Titus really did use the indexes as Pliny suggested (pref. 33), he might well have missed them. It was his *History* which Pliny intended to be the main vehicle of his loyal feelings (pref. 20). He implicitly contrasts it with the lowly subject-matter of the *HN*, which has little import or grandeur (pref. 12–14). In fact, the rhetorical ferocity which he lavishes on some of his Neronian passages may be connected with the themes of the *HN* itself. Nero, as depicted in the *HN*, epitomizes the antithesis of many Plinian ideals expressed therein. For example, he is not just luxurious, but shows the perverted ingenuity for innovations and refinements in luxury[46] which Pliny finds particularly degenerate.[47] Three

[45] Cic. *Att.* 2. 1. 8.
[46] *HN* 12. 83, 13. 22, 16. 233 (cf. Sen. *Ben.* 7. 9. 2), 34. 63.
[47] See Ch. II. 4.

passages all criticize Poppaea's gold-shod herd of mules, whose sole purpose is to provide her with enough milk to bathe in completely, rather than just using it on the face alone, as was the more usual practice (28. 183). Nero's use of medicine is connected not with *humana salus*,[48] but with immoral pleasures and over-indulgence. He uses a remedy for bruises incurred in night brawls (*HN* 13. 126), and drinks a concoction made from boars' dung in order to protect him in a chariot-race (28. 238). Above all, Pliny's description of Nero as *hostis generis humani* (7. 45–6) makes him the antithesis of the whole humanitarian ideal of the *HN* It sets him in opposition to those men who preserve or promote human life.[49] His interest in magic is linked with his *saevitas* towards the human race.[50] On a wider scale, his activities are reflected by discord rather than harmony in the natural order. The omens which foreshadow or accompany his death are particularly spectacular examples of the disruption and perversion of *Natura* herself.[51] Of relevance to this line of thought is a recent theory concerning the adverse effect Nero's activities had on contemporary society's attitude to *artes*.[52] Certainly, a survey of Pliny's comments in the *HN* on Nero suggests that his feelings about the corruption of *artes* by *luxuria* were given extra point by the contemporary actuality of Nero's extravagances. It is also true that, by contrast, he views with approval an attitude to *artes* which is more in line with traditional Roman *mores* and which is, moreover, specifically linked at times with Flavian ideology. There is an implicit contrast between the public building-schemes of Augustus and of Vespasian and the private luxury of the imperial palace of Gaius and the Domus Aurea of Nero.[53]

Typically Roman cultural values also appear, at first sight, to explain Pliny's attitude towards the Greeks. Generally, the omnivorous learning of the Greeks was considered to be combined with a total lack of virtue and propriety. They were

48 Ch. VI, intro.

49 e.g. 2. 18, 54. See Ch. II, *passim*.

50 *HN* 30. 14–15. See Ch. III. 2. B. ii.

51 *HN* 2. 199, 232. See also Ch. III. 1 and n. 22.

52 I. Lana, 'Scienza e politica in età imperiale romana', in *Tecnologia*, 21–43. See also the comments and modifications in M. Pani's review, *Quaderni di storia*, 18 (1983), 363–6, and now Isager (n. 25), 224–9.

53 *HN* 36. 102, 111–12. See also Ch. V. 2.D and nn. 58, 60.

wanting in dignity, seriousness, and respect for themselves and for others. They were deceitful and even mendacious, unscrupulous, and above all highly boastful about their cultural superiority to the Romans. Of the large number of references to aspects of these feelings in Latin literature, Pliny's quotation of Cato the Elder sums up much of the standard anti-Greek attitude (29. 14–15). 'They are a quite worthless people and an intractable one . . . When that race gives us its literature, it will corrupt all things and all the more if it sends hither its physicians. They have conspired together to murder all non-Greeks with their medicine and actually charge a fee to gain credibility and destroy us easily. They are also always calling us uncultivated foreigners . . .'.

On further examination, however, Pliny's own feelings appear more ambivalent. Certainly, a belief in the moral superiority of Rome continually asserts itself: 'The one race of outstanding eminence in virtue among all the races in the whole world is undoubtedly the Roman' (7. 130 ff.).[54] Yet his research owes a great deal to the products of Greek culture, as a glance at his indexes of sources will immediately show. Time and again he is forced to quote Greek writers;[55] even, reluctantly, in his description of Italy herself (3. 122). His assertion of Roman *virtus* in book 7 cannot disguise the predominance of the Greeks in such intellectual pursuits as poetry, philosophy, rhetoric, and drama (7. 107 ff.). In fact, it is noticeable that Pliny is particularly sensitive to the Greeks' intellectual claims, as would be expected in an author who aimed to outdo them by covering in one work all aspects of their *enkyklios paideia* (pref. 14). Unreliability and mendacity are frequently attributed to them, notably in Pliny's account of the nature and provenance of amber. Here, he makes unfavourable comments about their *fabulositas*, which is linked with the condemnatory term *vanitas*, also used in his attacks on another group of lying foreigners, the Magi (see below). Above all, he is galled by their boastful superiority: 'a people most prone to gushing self-praise' (3. 42). There is irony in the remark that Athens has no need of advertisement (*praeconii*, 4. 24) in Pliny's itinerary. Linked to this is their *loquacitas*, which is very

[54] Cf. also remarks on Roman *pietas, HN* 7. 121 ff.
[55] e.g. 2. 8, 13, 53.

evident in Pliny's criticism of the medical sects in book 29. In medicine, their tendency to frivolity in intellectual matters (cf. his remarks on fancy Greek titles to books, pref. 24) is particularly culpable. They are disrespectful of the serious matters of life and death, and their curiosity is not curbed by ethical consideration. They are happy, for instance, to dispense immoral (28. 5–6) or noxious (25. 24) remedies: 'medicine alone of the Greek arts has not yet exercised Roman *gravitas*' (29.17). As we shall see, Pliny (Ch. VI) has a personal preoccupation with Roman herbal remedies,[56] and this, of course, adds to his criticisms of Greek medicine. These and other medical passages, however, suggest that Pliny saw the Greeks as posing a threat more fundamental than that of literary rivalry alone to his ideas. By neglecting herbal medicine in favour of 'theories' or by making compounds out of simple remedies, their overingenious and unscrupulous minds are working against rather than in accordance with *Natura*.[57]

When Pliny denounces the Greeks for unscrupulous and immoral uses of human bodies in medicine (28. 5–7), he compares them to the Eastern exponents of magic, the Magi. Magic is denounced even more vociferously than the Greeks in the *HN*. The author can, on the one hand, be seen as expressing the strong but indefinable reactions of his contemporaries, best described in modern terms as a mixture of xenophobia, of common sense insulted, and of a revulsion from the nasty and unpleasant. As with the Greeks, his personal interest in a Roman 'herbal' medicine is also relevant. But he is aware, too, that magical marvels challenge or contradict his own consciously created ideal of the power of *Natura*, the origin of all life. It is, in Pliny's view, the Magi who more than anyone else act *contra naturam* and thus excite his most rhetorical denunciations.[58]

Finally, the marks of Pliny's personality are by no means

[56] See index s.v. 'herbals', esp. Ch. VI. 2. D.

[57] See J.P.V.D. Balsdon, *Romans and Aliens* (London, 1979), 32–3 on *loquacitas* and 31 on levitas. On Pliny's attitude to Greeks generally see G. Serbat, 'Il y a Grecs et Grecs! Quel sens donner au prétendu anti-hellénisme de Pline?', *Helm*. 38 (1987), 272–82. The philosophical as well as the intellectual misgivings of Pliny with regard to Greeks are most clearly laid out in A. Wallace-Hadrill, 'Pliny the Elder and Man's Unnatural History', *G&R* NS 37 (1990), 80–96. See also Ch. VI, *passim*, on medicine and n. 2 for general bibliography.

[58] See Index s.v. 'Magic', and '*Natura*, divine power of', esp. III. 2. A. ii.

confined to topics involving passages of invective. A more constructive theme like the discussion of agriculture is no rehash of Virgilian idealism or the practical precepts of Cato and Varro. It includes clear evidence of a personal interest in the latest developments, and its emphasis upon organization bears the unmistakable stamp of the military and administrative background of its author.[59]

These examples should offset the frequent criticism[60] that Pliny was a faceless compiler. He may have used a number of diverse sources, but these were subsumed into the clear thematic structure of the *HN*, as described earlier. Such creative collecting should be distinguished from the scrap-book technique of excerptions and abridgements found in later compilations and culminating in the tenth-century *Suda*. It was another characteristic of Pliny's age; it may be seen, for example, in his nephew's letters, which, unlike those of Cicero, were consciously composed and collected with a view to publication. The juristic works of Salvius Julianus under Hadrian exhibited a greater spirit of creative originality than is to be found in the sixth-century *Digest* of Justinian.

Proof of the creative intentions of the *HN's* author may also be found in Pliny's awareness of the significance of his work, in both an intellectual and a literary sense. For Pliny, the compilation of past knowledge is vital to future developments. But, while the pages of the *HN* contain much reverence for his predecessors, they also show equal interest on Pliny's part in what is new and only recently discovered. Occasionally, he will even claim to be the first to propound a particular piece of information.[61] As for literary awareness, Pliny's consciously rhetorical style has been ridiculed for its tortuousness,[62] but the significance of this consciousness should not be overlooked. Literary and intellectual awareness come together in his criticisms of the contemporary

[59] See Ch. V. 1, *passim*, and V. 1. B. iii on organizational skills.

[60] See e.g. H. J. Rose, *A Handbook of Latin Literature*[3] (London and New York, repr. 1966), 347.

[61] See Index s.v. 'progress', especially Ch. II. 1.

[62] E. Norden, *Die antike Kunstprosa* (Leipzig and Berlin, repr. 1909), 314, describes Pliny's style as one of the worst extant. J. Wight Duff, *A Literary History of Rome in the Silver Age*, ed. A. M. Duff (London, 1966), 307–9, talking of Pliny's 'fanciful style' and 'exasperating . . . maltreatment of the ablative case', concludes that 'fortunately, the world has not judged Pliny by his style'.

agricultural writer Columella[63] and in his self-conscious pioneering of the literary transmission of Roman herbal medicine.

'A work of great size and learning, as diverse as Nature herself' was the younger Pliny's description of his uncle's *HN*. The scope of the *HN* has given it a place in the culture of successive ages to a degree unparalleled by any comparable work of ancient literature. Its very survival is in itself a testament to its unique popularity: Varro's encyclopaedic work did not survive the Middle Ages; some medieval encyclopaedias were never printed, others were last reprinted in the seventeenth century. Such, for example, was the fate of the great *Speculum Mundi*, composed in four parts by the Dominican Vincent of Beauvais in about 1244.[64] On the other hand, it has been estimated that, by the early part of this century, no fewer than 222 editions of the *HN* had been published, to say nothing of 42 partial editions and 62 criticisms.[65]

In the two centuries following the death of Pliny, the *HN* is attested by his nephew, and by Suetonius and Gellius.[66] It was used extensively by later encyclopaedic writers, starting with Solinus at the beginning of the third century.[67] Of the early Christian fathers, Pliny's work is praised by Jerome, who lists the author as the Latin equivalent of Aristotle and Theophrastus.[68] Later, Bede had access to at least half the *HN*, which he termed *opus pulcherrimum*, and he may have helped popularize the work in Europe.[69]

However, tracing the influence of the *HN* on later writers soon runs into problems. From the second century, the *HN*, especially

[63] See Ch. V. 1. B. ii.

[64] Last published in 1624. See M. Chibnall, 'Pliny's *Natural History* and the Middle Ages', in T. A. Dorey (ed.), *Empire and Aftermath* (London, 1975), 57.

[65] E. W. Gudger, 'Pliny's *Historia Naturalis*: The Most Popular Natural History Ever Published', *Isis*, 6 (1924), 269–81.

[66] A. Labhardt, 'Quelques témoignages d'auteurs latins sur la personnalité et l'œuvre de Pline l'Ancien', in *Mélanges offerts à Max Niedermann* (Neuchâtel, 1944), 105–15.

[67] Grimal (n. 34); W. H. Stahl, *Martianus Capella and the Seven Liberal Arts* (New York, 1971), i. 47 ff.; J. Oroz Reta, 'Présence de Pline dans les Étymologies de saint Isidore de Seville', *Helm.* 38 (1987), 295–306.

[68] Jerome, *In Ezech.* 9. 28. 11 (PL 25. 271 c Migne); *Adv. Iovin.* 2. 6, (PL 23. 306 B Migne). See P. Courcelle, *Late Latin Authors and their Greek Sources*, tr. H. E. Wedeck (Cambridge, Mass., 1969).

[69] Bede, *De Temp. Rat.* 27 (PL 90. 412 Migne). See Chibnall (n. 64), 61–3, and P. Riché, *L'Éducation et culture dans l'Occident barbare* (Paris, 1962), 434.

the medical books, had been abridged and epitomized. Moreover, ancient authors did not follow modern conventions in citing their sources. Even if an author quotes Pliny by name, he may still have borrowed the complete passage from an intermediary source. Even some of Jerome's Plinian stories may have been derived from an intermediate source such as Tertullian.[70] Generally speaking, the more distant in time a writer is from Pliny, the less likely he is to have used the *HN* directly.[71] The subject of Pliny's later influence is too vast for a comprehensive treatment, though a large number of specialist studies have appeared in recent years.[72] On a more general level, it is possible to detect different approaches to the *HN* by readers in different ages.[73] For instance, early medieval readers scoured the *HN* for facts and anecdotes. With the Renaissance came a new interest in Pliny's literary and linguistic side. His work provided knowledge of lost classical authors and his scientific vocabulary helped in the translation of Greek scientific works. With the advent of printing came a new emphasis on the correctness of the published text and systematic textual restoration and commentary came to dominate Plinian studies. During the sixteenth century, however, the criticism of Pliny on the basis of science rather than philology gradually gained ground. In the later sixteenth century, the German botanist Melchior Guilandinus said that literary works could be emended by the use of codices only, but scientific ones must use other sources of information on the same topic. With the development of scientific knowledge, Pliny gradually lost his position of pre-eminence. By the twentieth century he was regarded as a mine of misinformation, if not actual fantasy.

It is only comparatively recently that the value of the *HN* as a historical document of first-century attitudes has been

[70] Courcelle (n. 68), 88–9.

[71] G. Sabbah, 'La référence comme indice de distance dans l'énoncé de Pline l'Ancien', *R.Ph.* 47 (1973), 38–49; Chibnall (n. 64), 61; G. Sabbah, 'Présence de la *HN* chez les auteurs de l'Antiquité tardive', *Helm.* 38 (1987), 203–21.

[72] See G. Serbat, 'Pline l'Ancien: état present des études sur sa vie, son œuvre et son influence', *ANRW* II. 32. 4, 2069–200. This bibliography concentrates on literature since 1970. For earlier studies, see K. Sallmann, 'Plinius der Ältere 1938–1970', *Lustrum*, 18 (1975), 5–299; 345–52.

[73] The following outline is based on C. G. Nauert Jnr., 'Humanists, Scientists and Pliny', *Am. Hist. Rev.* 84 (1979), 71–85. Cf. also R. French, 'Pliny and Renaissance Medicine', in *Science in the Early Empire*, 252–81.

acknowledged. The creativity of the *HN* and the persona[74] of its author have tended to be underestimated, but the situation is changing: it has, for example, been more frequently recognized that Pliny's concept of *Natura* is more than a superficial linking device in the *HN*,[75] and that a utilitarian and humanitarian attitude is a distinctive feature of his writing.[76] Yet, even now, his work is often regarded, first and foremost, as an unwieldy mass of raw material, with varying estimates of its reliability, for historians; a source for studies of particular topics, such as ancient medicine, art, or architecture.[77] Specialist inquiries are obviously of considerable interest and value, but they do not concern themselves with the *HN* for its own sake. Symbolically, a recently published collection of studies concedes only in its closing paragraphs that 'the result of specialist inquiries into the *Natural History* must necessarily be as diverse as the topics to which Pliny addressed himself, and lack of a cohering argument is a complaint that the commentators have raised against the *Natural History*. Yet overall there is a characteristic Plinian approach. . . .'[78] Now that the *HN* has outlived its usefulness as

[74] Now treated by G. Grüninger, *Untersuchungen zur Persönlichkeit des älteren Plinius: Die Bedeutung wissenschaftlicher Arbeit in seinem Denken* (diss. Freiburg, 1976). More recent articles are developing the idea of a coherent philosophy in Pliny: P. Grimal, 'Pline et les philosophes', *Helm.* 37 (1986), 239–49; M. Ducos, 'Existe-t-il une philosophie du droit chez Pline?', ibid. 201–18.

[75] R. Schilling, 'La place de Pline l'Ancien dans la littérature technique', *R.Ph.* 52 (1978), 272–83, comes nearer than most to giving the *Natura* theme the emphasis that is its due, cf. also G. Conte's introduction to his Italian edition and commentary, *Storia naturale*, i (bks. 1–6; Turin, 1982). More interest in this theme is now emerging: K Sallmann, 'La responsabilité de l'homme face à la Nature', *Helm.* 37 (1986), 251–66; Wallace-Hadrill (n. 37). Now also J. Isager's study of art history in Pliny (n. 25), which appeared as this book was going to press. He stresses the importance of viewing the art chapters in context and therefore precedes his examination of them with a discussion of Man and Nature in the *HN* (16–17, 32–42).

[76] e.g. S. Citroni Marchetti, '*Iuvare mortalem*: L'ideale programmatico della *NH* di Plinio nei rapporti com il moralismo stoico-diatribico', *A&R* 27 (1982), 124–48.

[77] A perusal of the bibliographies (n. 72) will serve to show that the vast majority of these are of a very specialized nature.

[78] *Science in the Early Empire*, 276. The existence of 'the Plinian approach' is recognized, but forms a background to, rather than the main point of, the studies. There are, however, reflections relevant to the present examination in O. Pedersen, 'Some Astronomical Topics in Pliny', 166–70 (Pliny's concept of the universe and practical view of astronomy), and in the conclusion of A. Locher, 'The Structure of the *Natural History*', 276 (Pliny's sensitivity to all aspects of the natural world). Interest is clearly growing, as may be seen in the more recent Nantes conference papers (*Helm.* 37 and 38), and the articles cited in nn. 74–6.

a practical handbook for everyday life, it is possible to step back to examine this 'Plinian approach' and view the *HN* as a whole. The result is a portrait not just of 'nature, that is, life', but also of the author himself and the age in which he lived.

I

Divina Natura: The Roots of Pliny's Thought

1. Natura *as Divinity*

THE world . . . is fitly believed to be a deity, eternal and immeasurable, a being that was never born and will never perish. What lies outside it does not concern men to explore and is beyond the capability of the human mind to guess. It is sacred, eternal, immeasurable, wholly within the whole, or rather itself the whole, finite and resembling the infinite, certain of all things and resembling the uncertain, holding in its embrace all things that are within and without, at once the work of nature and nature herself (*HN* 2. 1).

Pliny starts the first book proper of the *HN* in grand style. The scale of his chosen subject could not have been more emphatically described. *Natura* is the world, both as a whole and as its separate components; she is both the creator and the creation. The comprehensiveness of Pliny's analysis defies further elaboration: *Natura* is everything. Notwithstanding, long sections of commentaries[1] have been devoted to this and succeeding paragraphs of Pliny's introduction to the cosmos. But too detailed an exposition is of limited usefulness in the case of an eclectic and non-specialist author. Pliny's ideas need to be viewed more broadly in terms of general tradition, key ideas, and recurring patterns of thought.

A long tradition of Greek writing bore witness to the prime importance of thought on the nature of the cosmos,[2] not only from the famous philosophers like Plato and Aristotle, but also from more general writers as early as Hesiod. Pliny's opening

[1] Esp. W. Kroll, *Die Cosmologie des Plinius* (Breslau, 1930). Less fully, J. W. Caspar, *Roman Religion as Seen in Pliny's Natural History* (Chicago, 1934). Older editions are listed by J. Beaujeu in the Budé edition of bk. 2, pp. 115 ff. He also provides a succinct but useful commentary on these paragraphs.

[2] See S. Sambursky, *The Physical World of the Greeks* (London, 1963), 184 ff. The first exponent of a scientific cosmology was Anaximander. On Hesiod, see Philo, *De Aet. Mundi* 17.

words leave no doubt as to the main tradition which he is following. His divine, eternal, immeasurable, but finite *mundus* is directly derived from the cosmological theories of Plato, Aristotle, and the Stoics. While each of these had distinctive teachings, they all, broadly speaking, accepted the picture Pliny gives here. The main opponents of this tradition are given short shrift by Pliny. After briefly deriding as mad those 'who have taught that there are a countless number of worlds' (2. 3), he completely ignores[3] the atomists. The theories evolved principally by Democritus and followed most notably by Epicurus and Lucretius were opposed to most facets of the cosmology Pliny here espouses. The atomists believed that the world was perishable and, as Pliny's remark in *HN* 2. 3 suggests, that it was infinite. They also taught that it was mechanistic and denied the concepts of teleology and of a providential deity. Both these ideas were of central importance to Stoic cosmology generally and to Pliny's in particular (see sects. 3–4).

The atomists were, however, in a minority and destined to remain so for many centuries. The broad tradition Pliny adopts was in fact the prevalent view in the ancient world. Stoicism in particular, with its emphasis on duty in public as well as private life, was especially appealing to Romans. Yet Pliny's choice should not be seen as inevitable. At least one other Roman general writer with practical and scientific interests had, a century before, seen the world differently. Vitruvius' perception of architecture is coloured by an atomic view of the universe. Whether he is describing the proportions of the elements in building materials (*De Architectura* I. 2. 1 ff.) or the balance, symmetry, and proportions of the buildings themselves, the architect, perhaps appropriately, sees a mathematical, mechanical universe. In contrast, Pliny's interests centre upon human life. The Stoic cosmos was not only divine but also rational and intelligent. Stoic doctrine was also, by Pliny's day, becoming increasingly preoccupied with human issues (sect. 2). Against such a background, Pliny is often able to create an intimate and vivid picture of the relationship between man and nature.

[3] Works on magic (see Ch. III. 2. A) attributed to Democritus earn him Pliny's condemnation on several occasions (e.g. 24. 160; 25. 13; 26. 19; 27. 141). In 30. 9–10, he rejects the tradition that these writings were spurious. In 2. 14–15, Pliny claims that Democritus believed in only two gods, Punishment and Reward.

Within the framework of a mainstream tradition, then, Pliny brings to his portrait of *Natura* some more individual emphases. Even the most cursory reading of *HN* 2. 1–27 reveals one outstanding factor: the divinity of *Natura/Mundus*. Pliny's very first statement is that she is rightly believed to be divine. In 2. 13, he stresses the divinity of her *principale regimen*, the sun, also described as her soul and mind. Above all, more than half this section of book 2 (14–27) is taken up with a detailed discussion of the nature of God.

The trend of Pliny's discussion may be summed up as follows. To see God as having shape or form is a mark of human weakness. God is not a concrete entity. He consists entirely of self, sight, hearing, mind, and soul, a description with obvious similarities to that of *mundus* in 2. 1. By creating innumerable gods, human folly trivializes the concept of deity. Some of these human inventions ironically feature all too human failings, as the mythical stories of their loves and quarrels illustrate (2. 16–18). Others are assigned the ludicrously undignified forms of animals by foreign peoples. What matters is the relationship between man and man: 'Deus est mortali iuvare mortalem' (2. 18).[4] Man can be godlike by means of surpassing qualities rather than gods being human in the crudest sense. In essence, Pliny is expressing the Stoic ideal of service to one's fellow men, called by Seneca 'divina potentia' (*Clem.* 1. 26. 5).

At this point, Pliny digresses to dispose of two recently invented superstitions, astrology and the goddess *Fortuna*. In the case of *Fortuna*, he effectively stresses the paradox of placing faith in the personification of fickleness: 'Chance herself, by whom God is proved uncertain, takes the place of God' (2. 22–3).[5] The devotees of astrology adopt a parody of God as *certus* rather than *incertus* as in the case of *Fortuna*. Stoic philosophy included an element of determinism but God was still a vital power always at work.[6] In astrology, however, men

[4] A phrase probably derived from a commonplace (Caecil. 264 Ribbeck 'homo homini deus est, si suum officium sciat'; Cic. *Lig.* 38, see D. J. Campbell's edn. of *HN* 2 (Aberdeen, 1936), 53.

[5] On the rhetorical nature of this passage, see L. W. Rutland, '"Fortuna sola invocatur": Pliny's Statement', *Class. Bull.* 56 (1979), 28–31.

[6] See R. W. Sharples, *Alexander of Aphrodisias on Fate* (London, 1983), 8 ff. for a brief outline of the 'soft determinism' of the Stoic position. Chrysippus devised a compromise whereby the universe was governed by fate, yet room was

adopt an extreme fatalism and believe in a future predetermined to the extent of God's being idle: '. . . God's decree has been enacted once for all, while for the rest of time leisure has been granted to him' (ibid.).

What, then, is 'God'? For the next few paragraphs, Pliny continues to be negative in his search for a definition. The belief that God has a concern for human affairs is to be encouraged mainly for socio-political reasons and it stresses only the punishment of the wicked (2. 26). The good are rewarded, if at all, by their fellow men (2. 18).

Only in the final stages of the discussion does Pliny become more positive. His concluding remarks serve not only to define God but also to place man in a position of harmony and understanding with regard to him. Pliny admits man's *inperfecta natura* but says he is compensated for this by a knowledge that even God has limitations. These limitations are the inability to change the fact of his own existence, the course of history, or mathematical rules. For Pliny, these are proof of *naturae potentia*. That is, they are all laws of nature and he therefore concludes that Nature is what we mean by God.[7]

The whole passage raises a number of interesting points. Firstly, it highlights the extreme scepticism of Pliny with regard to popular religious beliefs (see Ch. III). Its importance in this respect seems to have escaped the attention of those who have

left for human responsibility: men are responsible for those actions which, though predestined, depend primarily on themselves. Determinism must not be an excuse for fatalism. Cf. *SVF* ii. 956; J. M. Rist, *Stoic Philosophy* (Cambridge, 1969), 112–32; S. Sambursky, *The Physics of the Stoics* (London, 1959), 49–80; F. H. Sandbach, *The Stoics* (London, 1975), 101–4; A. A. Long, *Problems in Stoicism* (London, 1971), 173–99. See also Ch. III. 1, *fin*, and n., 2. A. i, B n. 51).

[7] The actual examples are not as trivial or contrived as they appear at first glance. Each in its way expresses an idea of central importance to a Roman and a Stoic: (i) God cannot commit suicide. Suicide was God's greatest gift to men in Pliny's view (2. 27, cf. 28. 9, Ch. VI. 2. E) and that of Stoics generally. Pliny's example here thus enhances the ultimate freedom of the philosopher. (ii) God cannot alter men's acquisition of *honores*. Strictly speaking, a Stoic should regard *honores* with indifference. However, they were the obvious mark of success for a Roman performing his public duty. Pliny echoes the importance the Romans attached to the concept of *memoria*. The achievements of one man depend on the *memoria* of others in successive generations, rather than on any outside force. (iii) God cannot make 10 × 2 not equal to 20. Numbers were the basic bricks of the universe. Pliny says that *Natura* has to follow her own laws and is supreme precisely because they are unalterable. Is he once again thinking of a parallel with the Stoic, who was supreme if he lived 'according to nature'?

been content to label him a credulous writer. The significance of the mainly negative tone of the arguments will be examined shortly (sect. 3). Above all, his discussion illustrates well the difference between the exponent of Stoic philosophy and Pliny the writer with Stoic beliefs. In *HN* 2. 14–27 there is an allusiveness, a sense of unwritten assumptions, and a lack of strict clarity and coherence. This is more readily appreciated if the full text is read, rather than a summary. As suggested earlier, detailed philosophical criticism would serve no useful purpose here. Some background explanation and clarification is, however, clearly necessary if we are to gain a proper understanding of Pliny's views.

Overall, this view may be characterized as pantheistic. God is subsumed in the universe, rather than existing separately above it. Pliny talks of *naturae potentia* rather than *natura* in his final definition in *HN* 2. 27. While both Platonic and Stoic cosmological thought had mystic elements, the god of *Timaeus*, separate from and ruling over the universe, is clearly less relevant to Pliny than is the idea of a divine power permeating the universe, as generally held by the Stoics.[8] In *HN* 2. 208, Pliny himself talks of 'divine power of the nature which is diffused through everything' (*diffusae . . . per omne naturae . . . numen*) to which may be compared Chrysippus' view as recorded by Cicero that *vis divina* 'is to be found in reason (*ratio*) and in the soul and mind which pervades the whole of nature'.[9] The Stoic doctrine of *pneuma*[10] could encourage pantheistic ideas, and Philo of Alexandria (*De Aet. Mundi* 84) said that the Stoics believed God was the *psyche* of the world. Seneca suggests a pantheistic concept of God in the *Natural Questions*, when he says that God is all one can see (*NQ* 2. 45) and also all that one cannot see (*NQ* 1 pref. 13). Cicero's *De*

[8] *Timaeus*, esp. 28, 41 ff. Stoic pantheism: *SVF* ii. 1008–105, esp. 1022, 1026. The view that the whole universe is God is attributed to Chrysippus in Cic. *ND* 1.39.

[9] *ND* 1. 39, cf. *SVF* ii. 1025.

[10] *Pneuma*: *SVF* ii. 1028–48. See Sambursky, *Greeks* (n. 1), 133–4, where the history of the word's usage is given. The Stoics give active qualities to the fire and air making up *pneuma*; Sambursky stresses in particular the biological reason for this. Organic growth was thought to be inseparable from thermal processes. Thus, '*Natura* is a creative fire proceeding methodically to the work of generation' (*ND* 2. 57, attributed to Zeno).

Natura Deorum contains both concepts of God, but has several passages clearly identifying *mundus* and *deus* (e.g. 2. 46 and 2. 29) and, most importantly for comparisons with Pliny, a passage identifying God with both *mundus* and *natura*: '. . . deum esse mundum omnemque vim mundi natura divina contineri' (*ND* 2. 29, cf. Sen. *NQ* 2. 45).

However, the question of whether God was in or above the universe was still an open one, even for Stoics. At any rate, Seneca suggests as a suitable topic of philosophical speculation in *De Otio* 4. 2 'whether God encompasses his creation from outside or whether he pervades the whole of it'. The poet Manilius, whose ideas are mainly Stoic, is not consistent on this issue. He refers to a God of nature who permeates land, sea, and sky (*Astronomica* 2. 61) in language similar to Pliny's description of 'diffusae . . . per omne naturae . . . numen' (*HN* above, 2. 208). Yet, in other passages, he uses language more suited to the God of the *Timaeus*.[11] Seen against this background of uncertainty, the slight anomalies which occasionally appear in Pliny's pantheistic concept are not surprising and need not concern us. It is somewhat redundant, for example, in a pantheistic view of creation, to describe *sol* as *rector* of the universe.[12]

Another type of ambiguity in the philosophical conceptions of *natura/deus/mundus* may be discerned in the Stoic idea of *Natura artifex*, which is adopted by Pliny (e.g. 11. 1). This in effect makes her a deliberately creative power, rather than the impersonal, mechanical *physis* of atomists like Lucretius,[13] or the natural forces of the Peripatetic Strato of Lampsacus (*ND*

[11] Manilius, *Astr*. 2. 68: *Mundus* is a *machina* obedient to its overlord (*magistro*); 75–7: a world equal in its parts by the fairness of its creator; 81–2: over the whole *mundus*, things follow the guidance of a master (*dominum*), '. . . hic igitur deus et ratio quae cuncta gubernat . . .'.

[12] *Sol rector*: *HN* 2. 12–13, an idea perhaps to be linked with Plato, *Rep*. 6. 508. According to D.L. 7. 139, Cleanthes called the sun *to hegemonikon tou kosmou*, cf. Cic. *ND* 2. 40 ff. The Stoic emphasis on heat as life-giving (above, n. 10) and the sun's role in creating and preserving life may be of relevance here. For more speculative connections with Pythagorean ideas, see Beaujeu (n. 1), 125 n. 7.

[13] Lucretius does use the phrase 'natura creatrix (1. 629; 2. 1117; 5. 1362), but as no more than a figurative way of describing the impersonal mechanics of the atomic system. Bailey on 1. 629 suggests an analogy with Lucretius' equation of Venus with the creative power of Nature in the Prooemium.

1. 35).[14] Cicero had translated the great master-craftsman of the *Timaeus*, *demiourgos*, as *aedificator* and *opifex* (*ND* 1. 18). But he also showed (*ND* 2. 57) how the Stoic life-force, *pneuma* or *pyr technikon*, could become identified with *Natura*. His Stoic speaker, Balbus, says that Nature 'is a creative fire and the teacher of all other arts . . . but Nature in the universe as a whole is not an artist only but the master-artist . . . such is the nature of the *mens mundi*, so that it may be properly called the divine wisdom or providence, the Greeks' *Pronoia*, which has formed the world'. For the Stoic, the order and harmony achieved by *Natura* as *artifex* presupposed a conscious intelligence, *ratio* or *mens*. Far from being blind and accidental, she displays a benevolent purpose in making everything perfectly adapted to man's needs (*ND* 2. 81–167).[15] The benevolent providence of a rational deity is to be seen in Pliny's *Natura*,[16] and will be examined in sect. 3.

To conclude: the components are familiar, the picture Pliny builds with them is unique. The Stoic concept just described, combining a vague, impersonal pantheism with more personal notions of deity, coincides closely with the *Natura* described in the *HN*. The emphasis which Pliny lays on this pantheism is, however, unusual. In fact, the all-embracing general quality of his concept symbolizes the fundamental difference between his and Seneca's approaches to philosophy. Despite the appearance in his work of the pantheistic idea, Seneca often emphasizes a distinct heaven–earth divide. Describing the city of the world (*Ad Helv*. 8. 5), he says: 'no matter where you lift your gaze from earth to heaven, the realms of God and man are separated by an unattainable distance'. Unlike Pliny, Seneca the philosopher is concerned to stress the vital role of philosophy as the only means of bridging the gap between the human and the divine. For Pliny, no such void exists. The human and divine worlds are bound

[14] In Cic. *Acad*. 2. 121, Strato is said not to make use of the idea of divine activity in creating the world, but not to follow atomic theories either: creation is caused by natural forces of gravitation and motion.

[15] The idea of mind as the guiding principle of the universe appears first in fragments of Diogenes of Apollonia's *On Nature* which mention *noesis*. Better known is Anaxagoras' *nous*. See A. S. Pease, 'Caeli enarrant', *Harv. Theol. Review* 34 (1941), 163–200, esp. 164.

[16] *Natura*'s *providentia* and *benignitas*: e.g. 2. 154 (Mother Earth eulogy), 18. 5, 22. 1; see further sect. 3.

together. It is a relationship which forms the core of Pliny's thought and to which we now turn.

2. *Nature's Divinity, Man's Supremacy: Unity and Tension*

When the whole universe is vaguely divine, the position of man may be emphasized since there is no divine figure dwelling above to balance him. Conversely, Pliny's *divina natura* possesses the more personal qualities of the Stoic divinity which indicate a relationship with man. Her *providentia* and *benignitas*, impulses of an intelligent rather than mechanical being, are directed towards him. Pliny admires the *ingenium* with which she also provides plant cures even for the animals (22. 30). *Ratio* acts as a link between *Natura* and man, since man, too, has a share in it.[17]

It would, however, have been possible for Pliny to banish all ideas of divinity in the universe and concentrate on his human ideal, expressed clearly and succinctly in *HN* 2. 18: 'Deus est mortali iuvare mortalem'. Indeed, despite his deification of *Natura*, Pliny is often negative, and more of a sceptic than a Stoic in his approach to the question of God. In *HN* 2. 14–27, he arrives at the conclusion that *Natura* is God by process of elimination, ruling out a once-for-all creation which might imply a Platonic creator-god (2. 23), as well as the Olympian deities and more recent beliefs such as astrology (2. 17–20, 23). The 'invention' of the goddess *Fortuna* by man himself makes conjecture about God even more obscure as far as Pliny is concerned (2. 22). *Natura* is God—provided that there is a God at all (2. 14). Not only is Pliny's process of argument essentially negative; his definition of God as the whole of nature is so wide as to make it only one step from Plinian pantheism to atheism. Yet, while the discussion in *HN* 2. 14–27 is interesting for its attack on many traditional beliefs and superstitions (Ch. III. 1), it leads nevertheless to the conclusion that Nature is God (2. 27). In practice, this just means that Pliny can vary the stress on Nature's

[17] For *ratio* as expressing a relationship between man and God, see Sen. *NQ* 1, pref. 14: the higher part of man is *animus*, but God is nothing but *animus*. He is all *ratio*, cf. *Ep*. 66. 12, 'ratio autem nihil aliud est quam in corpus humanun pars divini spiritus mersa'. He goes on to ask 'Quid est ergo ratio? Naturae imitatio' (66. 9).

divinity according to his context. The overall result is that he achieves a remarkable harmony between the concept of *divina Natura* and a human interest which is concerned primarily with the everyday existence of man.

The pragmatism of Pliny's emphasis on man is to be seen, for example, in a difference of emphasis between his and Lucretius' concern for human fears. Pliny wishes to dispel fear as it is encountered in the day-to-day existence of man. To a certain extent, this is true of Lucretius also, but he is more concerned than is Pliny over the ultimate superstitious fear of an afterlife. To this end, he is at pains to dissociate the human world from all divine influence and advances a purely mechanistic view of the universe (see Ch. II. 3).

'Human interest' is often remarked upon as a characteristic of later Stoicism,[18] but even Seneca is largely concerned with moral values and with the ultimate fear of death, and advocates disregard of life generally.[19] Pliny, in contrast, gives the impression that he approves of practical comfort, of a life that is healthy and reasonably pleasant.[20] His interest in man need not be attributed solely to Stoic influence. There is much in his mode of thought to be compared with Protagoras and the early Sophists. Such a comparison must be a general one, rather than a specific comparison with the stricter meaning of Protagoras' 'man is the measure' saying.[21] The character of the Sophists' ideas—rational, humanist, and utilitarian—bears a general resemblance to Pliny's outlook.[22] Despite the divinity of *Natura*, man is, loosely

[18] Sen. *Clem.* 1. 3. 2: 'man is a social animal born for the common good'; *De Vit. Beat.* 24. 3, Nature bids us do good to all men, whether slave or free; 'wherever there is a human being, there is a chance to perform a service'. Cf. Marcus Aurelius, *Med.* 7. 55, 8. 59; Epictetus 3. 13. 3, 3. 24. 58, 4. 4. 27, fr. 7.

[19] Sen. *Ep.* 87. 36, on the difference between advantages and real goods. Goods admired by the crowd are not real goods (*Ep.* 66. 31) but are illusory and fleeting (*De Brev. Vit.* 17. 3). Life is said to be lamentable (*Ad Marc.* 11. 1) and is compared to slavery (*De Tranq. Animi* 10. 3). He advocates scorn of life (*De Tranq. Animi* 11. 4; *Ep.* 4. 4, 8), and postulates a celestial abode for the soul (*Ep.* 75. 18; 102. 22, 28; *Ad Marc.* 23. 1–3). In *NQ* 6. 32. 4–7, he argues against fear of death and regard for life: 'rerum natura te, quae genuit, exspectat et locus melior ac tutior' (6).

[20] See Chs. II. 4; V. 2. D on luxury, and VI, esp. 2. E, on health.

[21] See W. K. C. Guthrie, *A History of Greek Philosophy*, iii [Cambridge, 1969], 171 ff. See Intro., p. 13.

[22] Guthrie (n. 21), 14–26, 44–8. It is also interesting that Protagoras was a declared agnostic, an attitude which may be compared with Pliny's air of uncertainty in *HN* 2. 14 (. . .'Whoever God is—if there is a God . . .'). Protagoras'

speaking, very often the measure of all things for Pliny. 'Deus est mortali iuvare mortalem' suggests that moral virtue is very much a human concept. It is interesting in this respect that he rejects the deification of virtues (2. 14).[23] This idea could be seen as an extension of the Stoic notion that all good is part of the divine, and was approved by the Stoic speaker in Cicero's *ND* (2. 61). Another concept absent from Pliny's *Natura* is the idea of the 'goodness' of the universe. Although she is *provida* and *benigna*, she is never said to possess *virtus*. Cicero, on the other hand, had come closer to Plato's concept of goodness (*Timaeus* 30 A–B) when he said that *virtus* 'can show itself in man, but how much more readily in the universe', since nothing is more perfect, and the highest quality, *virtus*, has to exist in the absolute and perfect (*ND* 2. 38–9). This use of *virtus* does, of course, lay stress on the idea of perfection rather than morality.[24] However, Cicero indicates that he is quoting an earlier source, Chrysippus, for this Platonically derived idea. This amoral sense may have already faded a little by his, and certainly by Pliny's, time as later Stoic ideas concentrated more upon *virtus* as a moral quality primarily applicable to man.

By not assigning *virtus* to his cosmic deity, Pliny is able to keep the divine aspect of *Natura* unobtrusive and our attention focused on man's activities. The stress he lays on this activity might be compared with the Sophists' emphasis on the human evolution of society but, unlike them, Pliny is concerned to show the harmony rather than the opposition of man with nature.[25] This is not to say

practical and humanistic reaction led him to reject the contradictory speculations of the natural philosophers (Guthrie 186, and 51, noting Gorgias, *Hel.* 13). Pliny expresses a very similar scepticism, but on the contradictory speculations concerning the nature of God, rather than the make-up of the universe (2. 14–27), see Ch. III. 1 on his rhetorical use of the diversity and contradiction of these speculations. For the difference in tone between this scepticism, based in classical rationality, and the Christian dismissal of Greek cosmological theories, see below, sect. 5.

[23] For such deifications, see Beaujeu (n. 1), 127 n. 3. The only other non-Christian critic listed by A. S. Pease (ed. and comm.), *M. Tulli Ciceronis De Natura Deorum* (Cambridge, Mass., 1955), 693, is Lucian, *Deorum Concilium* 13.

[24] See Pease ad loc.

[25] Cf. the celebration of man's material progress attributed to Protagoras in Plato's dialogue of that name. The contrast with the Sophists' stress on man's opposition to nature is, once again, applicable in a general sense only. There is nothing to suggest that Pliny took a conscious position on anything as complex as the *nomos–physis* controversy of the 5th c. BC; see Guthrie (n. 21) 64–5, and Ch. 4.

that *Natura* is merely a neutral background for man's activities. The relationship between the two contains tension as well as harmony; it is complex and often ambiguous, but it is always close. The following chapters will explore particular aspects of human qualities and achievements and their relation to *Natura*. For the moment, attention will be focused on complexities of this relationship which have a particular bearing on the idea of *Natura* as divine.

3. Natura provida *or* Natura saeviens?

This partnership or opposition of man and nature needs to be put in the context of the ancient views on teleology. In a general sense, the term denotes the view that the universe was created for a purpose, whether the unconscious teleology of an impersonal nature or the working out of a divine plan. In a more particular sense it is applied especially to the idea that the universe and its parts are planned for the benefit of the highest creation, man. The centrality of man in the universe was a fundamental aspect of the Greek outlook on life from earliest times and exerted a major influence right through to the later Christian era.[26] Such a view was reinforced by the ancient idea of man as a microcosm of the universe,[27] which could be carried to physical comparisons. In the *Timaeus*, man's head is said to be constructed on the model of the universe's divine sphere (44 D). The creator *Natura* is said by Manilius (*Astr.* 3. 50–1) to unite the limbs (*membra*) of the universe into a single body.

The link between man and cosmic deity was enhanced by the Stoic identification of the all-pervading spirit of the universe with

[26] R. Renehan, 'The Greek Anthropocentric View of Man', *HSCP* 85 (1981), 239–59. Pease, 'Caeli enarrant' (n. 15), 172, stresses that Aristotle, even in his later works, still saw animals as created for the sake of man. See further, Ch. IV. 1. A, B *init*.

[27] In the more general sense, the microcosm idea is a means of envisaging the divine purpose in the make-up of all parts of creation. See Pease, 'Caeli enarrant' (n. 15), 164, on Xen. *Mem.* 1. 4. 2–14. It is supported by the Stoic doctrine of sympathy (Pease (n. 15) 177). For the interest in the smallest parts of creation, see Pliny, *HN* 11. 4; Sen. *Clem.* 1. 19. 4; Galen, *De Usu Part.* iii. 238 Kühn, and a Christian adaptation in Tert. *Adv. In Marc.* 1. 14. See further below, sect. 4.

man's highest attribute, mind or reason. As early as Plato there had been gradations in this attribute as it went down the animal to the vegetable kingdom (*Timaeus* 77). By assigning *ratio* only to man, the Stoics laid emphasis on his unique position as opposed to the rest of creation. As we shall see (Ch. IV), it is a division to which Pliny, despite his fascination with the animal kingdom, adheres.

Teleological stress on the supremacy of man in creation also complemented the humanitarian trend of later Stoicism which we have already noted as a characteristic of Pliny's thought. Marcus Aurelius was to adapt the teleological idea to the ideal of social obligation when he wrote that the lower things were made for the sake of the higher and the higher for one another (*Med.* 5. 30).

It is typical of Pliny that he makes no direct philosophical statements[28] about man's unique superiority and central position in the universe. Instead, he expresses it through the cumulative effect of the *HN*; that is, through the actual structure of the work and through the ideas emphasized within it. His main discussion of the human race in book 7 follows the books describing the world and its features but precedes those devoted to the rest of the animal kingdom (bks. 8–11). His fascination with human society and achievements is not, however, confined to a single book but is impressed upon the whole work. Above all, the centrality of man in Pliny's universe is expressed in the recurring theme of Nature's Providence towards the human race.

The quality of *providentia* in the universe is, as we saw (sect. 1) closely linked with that of a divine mind as the guiding force. This is clear from the wording of Lucretius' objection in *De Rerum Natura*: 'I would venture to assert . . . that nature was certainly not created for us by divine power: it is so full of imperfections' (5. 196–9). Thus it is, initially, surprising when Pliny, who does not share Lucretius' mechanistic view of the universe, still comments on limits and contradictions to the providential view and does not try to hide them. Savage Nature threatens the land with volcanoes (*HN* 2. 236); she condemns part of the world to be permanently dark and snowbound (4. 88); her jealousy allows the rough sea to encroach on the land (6. 1).

[28] Though note *HN* 2. 95 (links between the souls of men and the heavens established by the studies of Hipparchus, cf. sect. 4), *HN* 2. 22 (above, sect. 1), 27. 3 (Roman race compared to the sun, *rector* of the universe, Ch. V. 2. c).

But it is worth bearing in mind two things. Firstly, Pliny's interest is not directed *exclusively* at the human race. The undoubted emphasis on man in the *HN* is to some degree balanced by his broad-minded curiosity, which takes an interest in the whole of Nature. He wishes to make the *HN* as wide-ranging as possible, both in the general subject-matter covered (pref. 14) and also in the sense of giving a complete picture of Nature: 'It is a difficult task to give . . . Nature to all things and all her properties to Nature' (ibid. 15). He also sees the craftsmanship of Nature as much in her smallest and most insignificant works as in her greatest. With regard to insects, he says 'Nature is to be found in her entirety nowhere more than in her smallest creations' (11. 4). Other animals, besides man, can to a certain extent be seen as microcosms. It is this, together with the interest in the range of species in the overall totality of Nature, which helps to balance the stress on man's supremacy.

Secondly, Pliny sometimes makes elaborate excuses for Nature's apparent lapses in providence which serve only to underline his predominantly anthropocentric outlook. There was a Stoic tradition of exculpating Nature, for example by arguing that her evils were only apparent and were in fact beneficial to man. This is the theme Pliny picks up in his remarks about poisonous honey, where he says that *Natura* intends to make man more prudent and less greedy thereby (21. 77–8).

The discussion of providence in Cicero's *De Natura Deorum* gives an idea of other arguments with which Pliny would have been acquainted. The Sceptic Cotta argues against Stoic ideas of providence in book 3. Even if the Stoics say that man's abuse of *ratio* is his own sin, a real providence would have foreseen and provided against this (*ND* 3. 78). It is this problem which leads Pliny to remark that iron is the best and worst of things (*HN* 34. 138). It enables man to practise agriculture, but, in order to kill other men more quickly, he has, by inventing arrows, even made it fly, 'sceleratissima humani ingenii fraus'. The *culpa* is not *Natura*'s but man's inasmuch as he has perverted *Natura*'s intention and made iron fly (34. 138).[29] Therefore, he concludes, it is due to *naturae benignitas . . . eademque providentia* that,

[29] This idea that the fault lies not with Nature or her produce but with man's perversion of them recurs frequently in Pliny. See Ch. II. 2, 4.

because of rust, nothing is more mortal than the substance most hateful to mortality. It is typical of Pliny's layman's approach to philosophy that he is not interested in solving the problem posed by Cotta, but accepts it as an interesting paradox, 'cutting the knot' (see Ch. II. 2). Elsewhere, he appeals to practical compromise, the general rule. Thus, in *HN* 18. 5, he just points out that in *Natura* the bad things are far outweighed by the good ones, so no one need doubt *naturae maiestas*.[30] This is related to the manner in which his moral arguments make one evil cancel out another. As poisonous honey discourages men from greed, so do snakes discourage him from the impious greed of mining. The apparent nuisance to man, rust, prevents to some extent his perverted use of iron.

Occasionally, the providential view of Nature is replaced by a more impersonal notion of the natural balance of elements in *mundus* as a whole.[31] Thus, in *HN* 36. 1, Pliny says that, while everything he has described so far may be considered to have been created for man, *Natura* made mountains in order that she might curb her most restless elements, such as the violent force of heavy waves. Such comments are exceptional. These natural barriers are more often seen as *Natura*'s protection of man against the moral evil of his mistaken desire for gold and other luxury materials (e.g. 33. 3).

One part of *Natura* is always *provida* and *benigna* towards man in the *HN*. *Terra* is often personified and, in one passage, is even called *sacra parens* (33. 1–2), like *Natura* (e.g. 24. 1). Logically, his concept of *Natura* as both the material *mundus* and the spirit permeating it makes any deification of *Terra* redundant, but this, too, was a Stoic tendency, if less common than deification of *mundus* (*SVF*. ii. 92). Archidemus made it *hegemonikon* of the universe in much the same way that others portrayed the sun. Seneca suggests that the Stoics believed that the earth was a

[30] This citation of the need for an overall view may be compared to a passage of Maximus of Tyre (41. 4 Hobein), who said that, as with a human artist, it was necessary to look at Nature's work as a whole, not at individual defects.

[31] On the balance of Nature see the quotation of *SVF* ii. 528 in sect. 4 below. However, this goes on to say that the cosmos is 'the dwelling-place of gods and men, and the balance of those things created for their sake'. If man himself is the purpose for which the whole universe exists, there is little point in viewing him merely as one of the parts of the universe which must be kept in balance with the rest of Nature: the balance itself is teleological.

living body.[32] He develops an analogy between veins and arteries and the earth's inner channels for water and air (*NQ* 3. 15. 1). Pliny makes such a parallel even more vivid when he talks about the earth's *viscera* (33. 2). Italy (37. 202) is said to be *gravida*. The idea of the special divinity of the earth was no doubt encouraged by the commonly held theory that she formed the stable centre-point of the universe (2. 1). In the *Timaeus*, she is the first and oldest of the secondary gods created by the Demiurge (40 B–C). For the author of the *De Mundo* (written between *c*.50 BC and AD 100), she is imperishable because of her biological balance and harmony (397^{a}24–^{b}8) despite being the heavenly body that is furthest away from the God's seat in the ether and therefore feeble, incoherent, and confused (397^{b}30–3).

More generally, it was a commonplace that Earth was man's nurse and mother, providing all his real needs. Philo (*De Aet. Mundi* 63 ff.) eulogizes Earth as the Pandora who gives both benefit and pleasure to man. However commonplace, a beneficent *Terra* enables Pliny, consciously or not, to gloss over the problems of an imperfect providence. In a famous passage (7.1 ff.), probably derived from a similar one in Cicero's *Republic*,[33] Pliny wonders whether *Natura* is not a stepmother rather than a mother to man. In contrast, *Terra* is always *parens*, never *noverca*. As we shall see, her functions in this role are important in the consideration of Pliny's remarks on man's use of the land in farming or gardening (Ch. V. 1; Ch. II. 5). In naturalistic terms, she is man's proper element and as such often contrasts with the dangerous and untrustworthy sea (Ch. V intro.).

The main features of Pliny's idea of providence in *Natura*, especially as displayed in man's own element, *Terra*, are conveniently summed up in his descriptions of mining (33. 1 ff., 70 ff.). In *HN* 33. 1 Pliny uses vivid similes of earth gaping or trembling (*dehiscere . . . intremescere*) as her *viscera* are probed.[34] This stresses the impiety of man's behaviour towards the divine—*indignatione sacrae parentis* (sect. 4)—as well as the immorality of seeking the material of luxury (gold) or of violence (iron).[35] A

[32] See H. M. Hine (ed. and comm.), *Seneca Natural Questions, Book 2* (New York, 1981), 141–2.

[33] *Rep.* 3. 1, see Ch. II. 3 and nn.

[34] Cf. Sil. It. 1. 231–3: 'Sed scelerum causas operit deus: Astur avarus/visceribus lacerae telluris mergitur imis.'

[35] He ignores in this instance the uses of iron in agriculture (34. 138).

more worthy motive would have been the search for medicines. *Medicina metallica* was a recognized concept and various medicinal uses of minerals are described by Pliny in *HN* 34. 166 ff. Yet, as Pliny remarks ironically, such remedies are hardly ever the object of these excavations. In any case, all the essentials for life, including many medicines, can be found on the earth's surface. But earth does not limit herself to the provision of bare essentials. If man took only what lay to hand, without digging under earth's surface, 'how innocent, how happy, how luxurious (*delicata*), even, life might be!' (33. 3). Earth has concealed things precisely because they are harmful to man; that is, *Natura* has made certain provisions against man's tendency to pervert *ratio*.[36] The good things which man is meant to have are given willingly and easily (*larga facilisque*, 33. 2).

In *HN* 33. 70 ff., Pliny describes a particularly difficult method of mining. Man eventually overcomes *Natura* in an unholy conquest which he compares unfavourably even with the seeking of luxuries (pearls and purple dyes) from the dangerous sea.[37] 'Spectant victores ruinam naturae' is the terse embodiment of his moral indignation.[38] The mountains which protect *Natura* from man and man from himself have been destroyed. Yet Pliny cannot avoid his optimistic tendency to admire the human achievement, even in an activity which violently rejects *Naturae providentia*. It comes across in his vivid description of the difficult procedure in *HN* 33. 71–3,[39] as he follows the process step by

[36] However, she has not made it impossible, only more difficult; the objection of Cotta is not answered. Cf. Sen. *Ep*. 110. 9–10 on injurious things hidden by *Natura* below the earth.

[37] According to Pliny, an old senatorial decree prohibited mining in Italy (33. 78, cf. 37. 202). Even outside Italy, mining was, in at least one case, subject to restrictions: the censors decreed that *publicani* could not have more than 5,000 men working in the gold-mines at Victumulae in the territory of Vercellae in NW Cisalpine Gaul (33. 78). Pliny's tone suggests that the senatorial decree had a moral purpose. Restrictive practices of the type he describes at Victumulae were, apparently, rare (see P. R. Lewis and G. D. B. Jones, 'Roman Gold-Mining in North-West Spain', *JRS* 60 (1970), 169 ff., App. 1, p. 181 on 33. 70 ff.). The purpose is not clear. Some sort of economic balance is assumed in the article cited. It is also possible that there was a fear of a large concentration of slaves in this area. But an element of moral concern over the profits reaped by over-zealous exploitation of the mines by individual *publicani* need not be discounted.

[38] These *victores* are to be contrasted with the heroes who have conquered in 2. 54. Their probing of the heavens, unlike these men's probing of the earth, was righteous because it freed man by conquering his irrational fear. See Ch. II. 3.

[39] See intro., p. 4.

step, up to the final collapse of the rock. Moreover, Pliny's last thought on the matter adds an interesting pragmatic touch to his otherwise idealistic criticisms: 'Yet even now gold has not yet been found, nor did they positively know there was any there when they began to dig; the mere hope of obtaining their coveted object was sufficient reason for encountering such dangers and expense' (33. 73). Pliny implies that the trouble taken to obtain the substance is disproportionate when set against its moral worth. This is an idea which will gain significance in the interpretation of Pliny's attitude towards luxury and morality (Chs. II. 4, V. 2. D).

4. *Contemplation or Investigation?*

Equally significant is the attitude to Nature herself displayed by such investigations. There was a general feeling in the ancient world that it was permissible to try to understand but not to control Nature. The ambiguities of man's intrusions into and alterations of the natural order are, as we shall see, a major and recurring theme in Pliny.

The ambivalent attitude of the ancients towards natural inquiries had a religious element. This can be glimpsed, for example, four and a half centuries earlier in fifth-century Athens, at the notorious trial of the philosopher Socrates. Political hostility had undoubtedly played an important role in this attack. However, by charging him with impiety, Socrates' enemies presumably reckoned they could also arouse some religious prejudice against a free-thinking intellectual. They may have been encouraged by his talk of his admonitory 'divine voice'. But more significant in the present context is the fact that he apparently felt the need to dissociate himself from a theory of the natural philosopher Anaxagoras that the sun and moon were not divine but merely lumps of rock (Plato, *Apology* 26 D–E).

Doubt surrounds the confused tradition that Anaxagoras had himself been prosecuted earlier for impiety. The point remains that a certain amount of popular superstition about natural philosophy must have existed to make stories like this at all plausible.[40] The same may be said of the anecdote (Plutarch, *Pericles* 32. 2) about Diopeithes the diviner's decree allegedly aimed at blackening Pericles through his association with Anaxagoras. This allowed prosecution of anyone who 'did not believe in the gods or who *taught theories about what was in the sky*'. Here we may at least have an interesting indication of the religious attitudes of Diopeithes and his supporters.[41]

Prior to the trial of Socrates, popular feeling that intellectual probing of the natural world led to impiety provided a basis for a caricature of the philosopher in Aristophanes' *Clouds* (423 BC). His Socrates was not only an immoral rhetorician but also a godless intruder into the secrets of nature. Socrates referred directly to this portrait when sketching the general prejudice against his meddlesomeness (*periergia*, cf. *polypragmosyne*, 31 B). He insisted that his inquiries were limited to the moral sphere and that he had no interest in 'things below the earth and in the sky' (*Apol.* 19 B).

In the first centuries of our era, this Socratic rejection of such investigations as irrelevant to *arete* was enthusiastically endorsed by certain Christian writers. Calculations of the earth's shape and shadow, of the formation of eclipses or of the positions and numbers of the stars, were declared trivial and useless since they did not lead to the divine truth. *Polypragmosyne* and *periergia* should be controlled (Basil, *Hex.* 1. 3, 8). It was sufficient to trust the Bible's information and remember that Moses had no need for such calculations (*Hex.* 1. 7). Irrelevance rather than impiety seems to be the predominant objection. Yet there are also hints of religious hostility, and these occur even in those Christian writers most steeped in classical culture.[42] Too much interest in the futile subtleties of the philosopher can lead to heresy (*Hex.*

[40] K. Dover, 'The Freedom of the Intellectual in Greek Society', *Talanta* 7 (1976), 24–54, doubted whether this and other 'impiety' stories reflect *official* repression of intellectual freedom in democratic Athens; political and personal enmities featured prominently in their backgrounds. But more recently (*The Greeks and their Legacy* (Oxford, 1988), 158) he writes that he now attaches more weight to the mood of superstitious fear and less to the political aspects in the case of Socrates' trial.

[41] See Dover, art. cit. 39–40, *The Greek and their Legacy*, 146–7.

[42] Pointlessness of natural inquiries: see D. S. Wallace-Hadrill, *The Greek Patristic View of Nature* (Manchester, 1968), Ch. I. Clement of Alexandria and Origen were more favourable: natural science was worthy of study, but only as a

3. 8). His is the attitude of 'one who considers himself wiser than the revelations of the Spirit' (*Hex*. 9. 1). Here, Basil attacks human intellectual arrogance, which in these cases seems also to amount to impiety.

Does Pliny place any such limitations on natural inquiries and, if so, to what extent are they moral, religious, or intellectual? In particular, does the concept of Divine Nature in any way impair his scientific zeal? The answer is not simple. On several occasions, he expresses reservations with regard to clever calculations about the earth and the heavens. Hipparchus is criticized because he 'dared to give the number of the stars to posterity and to enumerate the heavenly bodies by name', a feat 'etiam deo improbam' (2. 95). Eratosthenes' attempt to measure the circumference of the earth is also 'improbum ausum', 'but achieved by such subtle reasoning that one is ashamed to be sceptical' (2. 247, cf. 2. 87). These philosophers have clearly gone 'too far', but in what sense? Which worries Pliny more, the fact that man has overstepped his intellectual capacity or the fact that, in so doing, he may have overstepped a religious boundary? In spite of the fact that Hipparchus' attempt was 'etiam deo improbam', he is none the less introduced by Pliny at the beginning of the passage (2. 95) as someone who 'can never be sufficiently praised' for his work generally. Moreover, this work is described in highly religious and philosophical terms: 'No one has done more to prove the relationship of the stars with man and that our spirits are a part of heaven.' Clearly then Hipparchus is not to be regarded as a habitual attacker of religious concepts of the universe.

In the present instance, Pliny's vocabulary suggests he thought Hipparchus overambitious intellectually. Certainly the word *improbus* in these passages should be compared with Pliny's general use of it elsewhere in the *HN* to denote something that is

preliminary step towards the final goal, knowledge of God. The student must quickly move on from it towards eternal things (Clem. *Strom*. 2. 2. 5. 1). On the cultured background of Basil, see J. M. Rist, 'Basil's "Neoplatonism": its Background and Nature', in P. Fedwick (ed.), *Basil of Caesarea, Christian, Humanist, Ascetic*, i (Toronto, 1981), 137–220. The complexity of the Christian response to natural philosophy must not be underestimated. It varies according to time and place and is complicated by more general feelings of hostility to classical culture, found, for instance, in later Latin writings, or in non-Greek Easterners like Tatian. On the hexaemeral literature and the general Christian attitude to pagan culture, see further below, sect. 5 and nn. 55–60.

'too large'. It may, for instance, be used of Nature's power, in the sense that it is too large to comprehend: 'What is more powerful than a magnet? . . . Where in nature will you find a greater immensity (*maior improbitas*, 36. 126)?' Elsewhere, he talks of the *potentia* and *improbae vires* of *Natura opifex* when he admires the miraculous *mora*, the fish which can stop a ship whatever the weather (*immensum potentiae occultae documentum*, 32. 1). In the case of the philosophers it is their temerity (*audacia*) which is *improba*, a combination which suggests excess in a bad sense. They have been overambitious in seeking to measure *potentia occulta* so precisely. Nature for Pliny is essentially immeasurable by human minds and to think otherwise is self-delusory and pretentious. Eratosthenes' subtlety was a borderline case, Pliny being uncertain whether to condemn or admire the results. At the other end of the scale from these cosmic investigations, the painstakingly minute measures of natural substances in certain medical compounds induce him to cry out against *ostentatio artis* (*medicinae*): 'Which of the gods in the name of Truth fixed these proportions? It was certainly beyond the capability of man!' (29. 25, see Ch. II. 2, refs. and n. 26).

In contrast, when Pliny himself offers explanations of Nature, he does so in very general terms. Regular winds, for instance, may be caused by continuous motion of the world, or the fluctuation of the 'life spirit' itself, or by air whipped up by the movements of the planets (2. 116). He 'investigates' only how such phenomena fit into a basic system of precepts governing the workings of the universe as a whole. He is content with the conclusion that the regularity of these winds proves that there is a law of nature at work (2. 121).[43] When he says (2. 116) that this law is not 'unknown' even if it is not yet fully known (*percognitum*), we suspect that he is referring to this knowledge of its existence and that 'full' knowledge will be limited to the choice of one of the possible general causes that he lists.

[43] Note that *lex naturae* is equivalent to *ratio naturae* in *HN* 2. 113, where the irregular movements are contrasted with the regular in that they are *fortuita* and occur *nulla . . . ratione naturae*. The regular movements are 'fatidica ex alto [cf. the divine nature of the stars from which they come], statisque de causis et ex suis venire sideribus'. The contrast between *fortuita* and *ratione naturae* is interesting, since it can be related to the difference between *ratio* and *casus*, discussed in Ch. II. 2.

Rather than condemning the principle of discovery, Pliny is concerned that some claims to truth are misleading. Yet his intellectual misgivings cannot be divorced from religious ones. Pliny's vague, matter-of-course references to 'deity' in the Hipparchus and medical passages quoted earlier, point to two things: to the prevalence in ancient thought of a religious outlook on the world and to the consequent impossibility of assigning attitudes exclusively 'intellectual' or 'religious' labels. In the case of Pliny himself, while overt mention of the divinity of *Natura* is not to be found in every chapter of the *HN*, it underlies the work as a whole. As the ensuing chapters will illustrate, the interaction of man and nature is to an extent the interaction of man and the divine. To overstep the bounds of human *ratio* is to challenge the supreme *ratio divinae naturae* (Ch. II. 2). It is not merely intellectually unsound but also an implicit attempt to devalue the divine power of Nature. We shall also see that human motivation is of critical importance to Pliny's assessment of man's investigations into Nature. The philosophers Pliny criticized were overambitious but did not act in a spirit of impious attack on *divina Natura*. The fruits of their researches can often be put to admirable humanitarian uses as Pliny himself recognizes with approval in the case of eclipses (Ch. II. 3). The intellectual inquiries which Pliny does unambiguously condemn are those of the magicians. The concept of control rather than understanding is perhaps important here. The magician aims to control the power of Nature. Yet even in these instances, as we shall see (Ch. III. 2. A. i–iv), the condemnation is not solely for 'impiety' towards Nature. Rather, Pliny abhors their intentions towards humanity. In contrast to the predictor of eclipses, the magician aims to harm rather than to help other men; that is, to hinder rather than aid *vita* (Ch. II. 1), to disrupt rather than to cement the bond between man's life-force and its ultimate source.

There is another, more straightforwardly religious, aspect to man's understanding of Nature which influences Pliny. For the Stoics, contemplation of the beauty and artistry of the world proved there was a divine creator in or above the universe (Cic. *ND* 2. 87 ff.), an aspect of thought which comes close to the Christian one of contemplating God through his works: 'We are not born in order to see what has been created but in order that we may contemplate the creator himself, that is that we may see him with our mind' (Lact. *Inst.* 3. 9. 13).[44] The tradition is also to be found in Jewish writings.[45] Rather than seeking to improve on Nature, man is invited to stand back and admire the truly perfect. The Stoic's contemplation of Nature was one aspect of living in harmony with it. It was a means of fulfilling the individual's *ratio* and identifying with the divine. Cicero says (*ND* 2. 37): 'Man was himself born to contemplate and imitate the universe'.[46] The imitation idea has links with that of man as a microcosm.[47]

Expressions of awe and wonder at the greatness and majesty of Nature as displayed in her works are found throughout the *HN*. In *HN* 7. 7, for example, Pliny claims that it will not be possible to appreciate *naturae vis atque maiestas* fully if the mind takes in only specific details without grasping the whole picture. What makes his view rather more unusual, however, is the stress he lays on the contribution made by the smallest and humblest creations (sect. 3) to this grand overall picture. The divine creator is praised for the insect, one of her most wonderful efforts: 'Nowhere is there to be found a more striking example of her artistry' (11. 1–2). A Stoic definition of *mundus* refers to it as a balance of 'heaven, air, earth . . . and the sea and all the living things in them' (*SVF* ii. 528). However, later Stoics tended to stress the contemplation of only the most exalted aspects of nature,[48] such as the heavenly phenomena, and this is reflected in Seneca's distinction in the *NQ* 2. 3–4 between *partes* which are essential to the wholeness of *mundus* and *quasi partes* which are

[44] However, Christian writers were careful to guard against ideas of pantheism, which such a link between the spiritual and material worlds might induce, emphasizing that the ultimate object of admiration was the creator not the creation. See Wallace-Hadrill (n. 42) 127 ff. on this and on their use of facets of nature as symbols or anagogues of the spiritual world.

[45] Pease, 'Caeli enarrant' (n. 15), 190–1 and nn. 212a, 214, gives as examples the Nineteenth Psalm and the Wisdom of Solomon, though pointing out that these, like the later Christian writings, may have been influenced by Greek thought.

[46] See Pease on *ND* 2. 37. As an aspect of living in harmony with Nature, see Sen. *De Otio* 5. 8. 1: '. . . ergo secundum naturam vivo si totum me illi dedi, si illius admirator cultorque sum. Natura autem utrumque facere me voluit, et agere et contemplationi vacare.'

[47] Plato had already suggested (*Timaeus* 90 C–D) that contemplation of the revolutions and harmonious circuits of the universe should be practised, as these motions were akin to the divine element in man. In this way, man would more easily attain to truth and immortality.

[48] Cf. Sen. *NQ* 1, pref. 3, 1–3, 12–13: the more exalted the object of contemplation, the nearer it brings the viewer to the divine Mind.

not. The whole of the animal and vegetable kingdoms between them constitute a single *pars*. This implies that separate species, as well as individual animals or plants, are dispensable *quasi partes*.[49] In contrast, Pliny is more interested in all aspects of the natural world for their own sake. If some of his defences of his humble subject are made with literary prejudices in mind,[50] he is also aware of such philosophical ones: in *HN* 11.4, he criticizes both types of prejudice: 'I beg my readers not to let their contempt for many of these creatures lead them also to despise what I relate about them, *since in the contemplation of Nature nothing can possibly be considered superfluous*.'

This contrast between Pliny and Seneca with regard to the parts of Nature each considers worth studying is paralleled by their respective views concerning the extent of her divine element. Unlike the vague and variable pantheist Pliny, Seneca at the beginning of the *NQ* makes a clear distinction between study of the obvious parts of nature and that of the loftier mysteries of God and the secrets of the universe. For Seneca, man's *animus*, 'pars (mundi) magnificentissima' (*Ad Helv*. 8.4), can start by contemplating earthly things,[51] but only as a means to progressing to the divine celestial sphere, where it contemplates an eternity of time and space (*Ad Helv*. 20.2; *Ad Marc*. 26.5). Eventually, sense-perception should be abandoned for pure meditation: in *De Vita Beata* (8.4), he says that reason may at first investigate external things at the instigation of the senses, but it should eventually fall back on itself, for the soul is 'melius et certius lumen' than the eyes (*Vit. Beat*. 2.2). Any limitations are due to the restrictions of ability caused by human inferiority to the divine: only after death can the soul learn *arcana naturae* not *ex coniectura* but *ex vero* (*Ad Marc*. 25.2). Seneca's exaltation of contemplation is sometimes enhanced by language sug-

[49] The ambiguity of *physeis* (living things) in passages like *SVF* ii.528 just quoted is noted by Hine (n. 32), 164–71, on *NQ* 2.3–4. It can refer to individuals or to species. In either case, Pliny is closer to this than he is to Seneca's definition.

[50] e.g. pref. 12–14, 18.5. See Ch. V.1.B.ii.

[51] Earthly phenomena are not always unappreciated. In *Ad Marc*. 18.2ff., Seneca suggests turning to the spectacle of earth only after the viewer has been sated with the celestial spectacle. Yet he lists cities, plains, mountains, animals, cornfields, etc. with enthusiasm, and does not explicitly say that they are inferior, only 'alia forma rerum, aliterque mirabilis'. For Pliny, such sights supply man with pleasure and entertainment, as well as evidence of *Natura*'s grandeur. (See index s.vv. 'landscape' and '*spectaculum*').

gestive of religious awe: in *De Otio* 5. 8, he describes himself as nature's admirer and devotee (*cultor*). However, he retains a clear division between the human and divine worlds: there is a philosophy which deals with man and a philosophy which deals with God (*NQ* 1. 1. 1). The latter is 'higher . . . and more spiritual . . . and not embraced by the eyes'. That dealing with man is *publica*, while the other is *secretiora* (*NQ* 1. 1. 3). Pliny sees no such division.

A Senecan simile offers an appropriate conclusion. In *NQ* 7. 30. 6, he compares the higher philosophical approach to Nature to the more esoteric revelations associated with the mystery religions: 'Nature does not reveal all her mysteries at once, just as Eleusis keeps something in reserve to show those who revisit there . . . These secrets are not open to all indiscriminately. They are withdrawn and closed up in the inner sanctum.' These words suggest the intellectual exclusiveness of such philosophy, a contrast to Pliny's layman's approach. His *Natura* is, in comparison, more akin to those traditional gods whose wanton behaviour he complains about (2. 17). Although her benevolent side, described in the previous section, is the one most often seen, there is another, darker side, with more than a touch of the heartless caprice and cruelty of the Olympians, or even of the despised 'man-made' goddess Fortuna (2. 22). Pliny's *natura ludens* or *lasciviens* is not a cosy image; she revels in her own power, the luxuriant abundance of spontaneous growth (Ch. II. 5). When Pliny describes her creation of human freaks and oddities as 'ludibria sibi, nobis miracula' (7. 32), the power of the *Natura* divinity is shown to have rather a cruel streak. Again, while many of the things harmful to man in *Natura* are either hidden blessings, or become harmful through his perversion of them (see above, sect. 3), there is sometimes no reason for *Natura*'s wild savagery. Eruptions are a good example. When Etna erupts, '*natura* savagely threatens the lands with fire' (2. 236). Such upheavals of the normally beneficent Earth are 'scelera naturae' (*HN* 2. 206). Productive or destructive, Pliny's *Natura* is wild.[52]

[52] See Ch. II. 5 on cultivation and Ch. V intro. and 2. E on the sea as an aspect of *Natura*'s savagery. The existence of a wild, irrational element in *Natura* is suggested also by a passage in the *Timaeus* (48 A) in which it is said that the world is not the product of Reason alone, but of the combination of Reason and Necessity. Reason controlled Necessity by persuading it *for the most part* to bring

Pliny was right not to attribute 'goodness' to *Natura* (above, sect. 3): amorality is more appropriate. Yet an element of unpredictable savagery enhances rather than detracts from the power and majesty of an otherwise beneficent and providential deity.

5. *Postscript: Pliny's Cosmology in Context*

We have seen that Pliny built up his portrait of *divina Natura* by drawing on a long and rich tradition of cosmological theories and speculations in the ancient world. The result is a complex and ambivalent *Natura* whose characteristics often seem contradictory. The benevolent deity who serves the interests of her supreme creation, man, can sometimes unleash a chaos of unruly elements which threaten his very existence. On some occasions, she is no more than a backdrop for the works and deeds of man. At other times, her power imposes limitations on the ambitions of man's inferior intellect. She is stressed as the creator of mankind but also as the creator of the insect. Pliny encompasses the totality of Nature, investigating animal, vegetable, or mineral components in scientific detail.

It is not easy to interpret this in terms of the history of ancient thought. The ambiguity in Pliny's outlook has somethimes been elevated to the status of a key point[53] in the history of ancient science. He is held to embody a tension between two strains of thought; the Aristotelian and Presocratic analysis of precise phenomena, and the philosophical idea of sympathy and old Socratic prejudice against investigation unless it is aimed at *arete*. This places him between a more rational, analytical age and a less rational, increasingly superstitious, era that is developing. Pliny certainly displays different strands of thought in his work. However, it is unlikely that he was aware of such tensions. His acquisition of philosophy was more a matter of absorption than conscious critical analysis. He just was not interested in solving philosophical problems (see sect. 3). As for his chronological significance, it is, as has often been observed, difficult to iso-

about the best result. This does indeed imply that, in certain areas, reason may not be uppermost and that the confusion of the primeval chaos may still exist. See also Ch. II n. 37 on *rabies* in man the microcosm.

[53] O. Gigon, 'Pline', in *Plinio il Vecchio*, 41–52.

late points of greater or lesser rationality in a development so obviously complex and non-linear as ancient scientific thought. After all, the great physician Galen wrote some hundred years after Pliny, but he can hardly be called less 'analytical' or more 'superstitious' than his encyclopaedist predecessor.

It may be less precise but more rewarding to approach the question from a slightly different angle and take notice of Pliny's own testimony in the *HN*. He sees himself as the inheritor of the wisdom of the past in the broadest sense. He follows the example of all his predecessors who have assimilated and handed knowledge down to successive generations. He is, in short, an omnivorous inheritor of the classical tradition. Comments in the *HN*[54] show his awareness of this tradition. He is acutely conscious of the debt owed to men of earlier times, both for specific scientific discoveries and for the handing down of knowledge. He is anxious both for the advance of the former and for the continuation of the latter, an attitude belonging to the classical tradition of inquiry in general.

While keeping to such broad definitions, we can, none the less, further clarify Pliny's historical context. This classical attitude towards his intellectual heritage evokes an obvious contrast with the Christian outlook as it developed in the first centuries of our era. The major Church Fathers were well grounded in classical ideas and classical philosophy was adapted to the exposition of Christian belief.[55] Yet the actual recognition and appreciation of the classical tradition is conspicuously missing in official Christian writings, whatever the author's private view.[56] In the context of

[54] *HN* 2. 54, 2. 62, 14. 1; see Ch. II. 1.

[55] See F. E. Robbins, *The Hexaemeral Literature* (Chicago, 1912), Ch. 1. Complex biblical argument might be based on acquaintance with the Greek philosophical tradition. See Robbins 55 ff. on the detailed discussion of whether the fires of heaven destroy the waters below them in Gregory of Nyssa's *Hexaemeron*. He takes issue with Basil's view and makes use to a certain extent of Aristotelian theories. This learned Christian adaptation of pagan philosophy may in some respects be contrasted less with the classical tradition than with more unscientific and credulous Christian writers who discussed 'less important and even fantastic questions as suggested by the biblical text' (Robbins 59–63 on the school of Theodorus of Mopsuhestia). A pre-eminent example of the Christian learned in classical philosophy is, of course, Origen. On the background to his *Contra Celsum* see the introduction to H. Chadwick's translation and commentary (Cambridge, 1965), pp. ix–xxii.

[56] On the complexities of the Christian attitude to classical learning see W. Jaeger, *Early Christianity and Greek Paideia* (Cambridge, Mass., 1962), cf. N. G. Wilson, *St. Basil on Greek Literature* (London, 1975), intro. 9–11.

cosmological thought this is most appropriately illustrated by the *hexaemera*, or commentaries on the six days of the Mosaic account of creation. Like the classical cosmologies mentioned earlier in this chapter, those of Plato, Philo, or Cicero, they derived much from teleological thought.[57] With them, they create a continuous tradition of cosmological writing before and after Pliny's time. It is their motivation that is different: the upholding of faith is of paramount importance.

A glance at the first extant *hexaemeron*,[58] written in the fourth century by Basil of Caesarea, soon reveals modes of thought alien to Pliny. In a number of passages, philosophers are abused, misinterpreted, and dismissed in favour of biblical tradition and ideals of faith. Faith is, in fact, all too often the enemy of reasoned argument. In the second century, when Christianity was expanding, pagan writers such as Galen and Celsus criticized the Christian writings on precisely these grounds.[59] Pliny, writing in an earlier era, before this expansion, would be even less likely to understand the Christian ideal of faith and its claim to a monopoly of the truth. Basil contrasts the unartificial utterances of the Bible with the philosophers' forced reasonings: 'let the simplicity of faith be stronger than the deductions of reason' (*Hex.* 1. 10).[60] He dismisses them as liars, and the argument that they all contradict one another acquires a conclusive finality in Christian argument which is far removed from the scepticism of Cotta in the *De Natura Deorum* and from the uncertainty on the nature of God as expressed by Pliny himself in *HN* 2. 14–27. A similar stress on the primacy of faith as opposed to the con-

[57] Pease, 'Caeli enarrant' (n. 15), 194, points to the continuity of the Christian hexaemeral tradition with that of Philo. The *De Op. Mundi* combined Mosaic and Greek ideas from a Jewish viewpoint, while the later authors did the same from the Christian viewpoint. On teleology, see Ch. I. 3.

[58] The most influential of these works, widely imitated and adapted by later writers. Robbins (n. 55), Ch. 4; Pease, 'Caeli enarrant' (n. 15), 194 ff.

[59] Jaeger (n. 56), 32–3 and nn., cf. Robbins (n. 55), 46–7: Basil 'believed that matter was made out of nothing by God, but rather as an article of faith dependent upon the revelation to Moses, than as a thing logically to be proved'. Summary dismissals of philosophical theories which conflict with articles of faith may be seen in other Christian writings. Combating of heresy was, of course, an important consideration in the Christian emphasis on faith, e.g. Ambrose, *Hex.* 2. 8. 30.

[60] Basil, *Hex.* 1. 4, 3. 8: see sect. 4 and n. 42, and E. Amand de Mendieta, 'The Official Attitude of Basil of Caesarea as a Christian Bishop towards Greek Philosophy and Science', *Studies in Church History*, 13 (1976), 25–49.

tentious, oversubtle, and contradictory pagan disputations is to be found in Ambrose's adaptation of Basil's *Hexaemeron* (2.7.25; 3.2.5; 3.27). Even Basil's condemnation of the theory of the music of the spheres, as being useless for the Christian concept of God, is in a very different vein from Pliny's remark that it is *iucunda magis quam necessaria* (2.84).

We must, then, assign Pliny a place in the history of thought on the general grounds of attitude and tradition. We cannot assign him a more positive and specific intellectual role. The development of thought is gradual and complex and not to be encapsulated in the writings of any one author, especially when he is a generalist rather than a philosopher, an inheritor rather than an innovator. Yet perhaps we can still ask whether there is any mode of thought which can place Pliny's writings in the first century, as opposed to the obvious echoes of earlier thought, and the supposed indications of future developments. In this case, it is his emphasis on humanity and its needs and aspirations which can most fairly be described as a general first-century phenomenon: it is found both in pagan philosophy—later Stoicism—and in Christian writings of this era and later. Of course, ideas on the supremacy of man were not in themselves new. Yet it is interesting that from this time, *mundus/kosmos* seems to broaden in meaning and more often denotes not just the world as the natural, physical structure of the universe, but especially the inhabited world. Longinus, in a well-known passage, talks of a 'cosmic' decline of literature (44.1). The Gospels, particularly that of John, and then patristic literature, produce a number of meanings, all of which refer to the cosmos as inhabited by man; 'mankind', 'the inhabitants of the world', 'the sphere of man's activity', 'the scene of life'.[61] Pliny himself indicates this meaning of the term in the contents of his *Historia Naturalis*. Long sections are devoted to man's arts as well as to the natural materials used; the uses of plants are given more space than descriptions of

[61] John 1:10; 3:17; 7:7; 12:13; 14:17, 22; 15:18, 19; 17:14, 21. Patristic literature: see *A Patristic Greek Lexicon*, ed. G. W. H. Lampe (Oxford, 1961). Mankind/inhabitants of the world: 1 Clem. 7.4; Origen, *Comm. in Mt.* 13.20. Sphere of activity: Dam. Papa, *Ep. Illyr.* ap. Thdt. *HE* 2.22.5. Scene of life: 1 Clem. 38.3; Arist. *Apol.* 1.1. In patristic literature, the word *kosmikos* often comes to mean worldly as opposed to spiritual and as such is rather ambiguous: see Gregory of Nyssa, *Traité de la Virginité*, ed. M. Aubineau (Sources Chrétiennes, 119; Paris, 1968), 480.

the plants themselves. Man and his cities are as much a part of *mundus* as are rivers and islands (7. 1).[62] Such an emphasis once again underlines the importance of the linking theme in this and the following chapters: the relationship between man and Nature in Pliny.

[62] 'Mundus . . . et in eo terrae, *gentes*, maria, flumina insignia, insulae, *urbes* . . .'. That man was a part of the cosmos was part of the classical cosmological tradition, cf. [Arist.] *De Mundo* 2, 391^{b}9–10; 3, 392^{b}19–20. It is the overall importance of the human element in Pliny which is remarkable. Cicero (*ND* 2. 88) uses the idea of man and his works being part of *mundus*, but it is in the context of a wider argument to prove that the universe must be guided by an intelligent force if some of its elements are in themselves rational and creative. Note also Seneca's use of the theme, emphasizing an interest in the heaven–earth divide which Pliny ignores (see above, n. 51). He describes man's entry into *tota vita*, the city of gods and men, lists the contents of the heavens, and follows with those of the earth, which include *urbes* and *nationes* (*Ad Marc.* 18. 2. ff.).

II

Man in Nature

Introduction

Despite the limitations on teleology described in the last chapter, man's central position in Pliny's view of the cosmos remains in no doubt. The works of man as well as of *Natura* may be included in a description of *mundus* and it is the sculpture, paintings, and buildings of man rather than the natural materials themselves that dominate the last five books of the *HN* (Ch. I. 5). Man emerges as a partner of *Natura*; his works are worthy to be set beside those of the great *artifex*. Indeed, if all the public buildings of Rome were brought together in one place, it would seem that *mundus alius* had been created by man (36. 101).[1]

'My subject is nature, that is, life' (pref. 13). This is life in the world generally, but especially that of man. We have seen that Pliny does not divide 'life', as Seneca does, into the higher philosophical one of the elevated and purified spirit and the lower one enslaved to the material world and the human weaknesses from which the philosopher has freed himself (Ch. I. 2; 4). Pliny's moral criticisms (e.g. 9. 67, 33. 39–41) may be very forceful, but are not allowed to obscure the overall tone of practical humanitarianism: his idea of life is a comfortable one (below, sect. 4). He himself stresses in the preface (12–13) that he is not confining himself to the most exalted aspects of human endeavour: the *orationes sermonesve* of the historical accounts of brave and famous deeds will not be found in his account of life. However, the *varietas* of *Natura* allows him plenty of opportunities for 'digressions . . . marvellous incidents or unusual happenings'. He apologizes for the need to stress the 'barren material' and 'least

[1] S. Citroni Marchetti, '*Iuvare mortalem*: L'ideale programmatico della *NH* di Plinio nei rapporti con il moralismo stoico-diatribico', *A&R* 27 (1982), 124–48, aptly compares this statement with that of Balbus in *ND* 2. 152 that man can build 'in rerum natura quasi alteram naturam'. See Ch. V. 2. c (*fin*) for comparisons of Rome/Italy and *Natura*.

elevated aspect' of *Natura*; such topics were considered to be beneath the dignity of serious literature. Yet their inclusion serves to stress the width of Pliny's vision of life. In fact, everything that can be said to promote life is there. The humble plants that can save a man's life are called into action far more often than the bravery of the individual, as epitomized by the unquenchable valour of M. Sergius Silus (7. 104 ff., see Ch. III. 1, 2. c).[2] Accordingly, they are awarded far more discussion. Pliny thus seeks to depict life in all its aspects. The stress on the promotion of life also suggests that his attitude is positive as well as humanitarian. *Vitae prodest*, *utilitas vitae*, and variants on these phrases[3] occur frequently: 'deus est mortali iuvare mortalem' (2. 18) sums up his attitude. All life's activities, inventions, and discoveries allow man to help other men, both contemporaries and descendants (see below, sect. 2). We would, therefore, expect him to have an optimistic view of man's progress and this will be the first topic in this examination of man's life in Pliny's Nature. Indeed, Pliny would be most unlikely to portray a life that was in any sense static. If his days were devoted to official service, his nights were devoted to study: to be asleep was not to be living (pref. 18–19). The passage corroborates his nephew's evidence (*Ep*. 3. 5) that even Pliny's hours of *otium* were the most active ones possible.

Man's place in *Natura* is not without ambiguities. His progress in some fields is offset by a moral decline which in turn reduces the desire to acquire and pass on knowledge both theoretical and practical (2. 117, 14. 1 ff, see sect. 1). His *ratio* is man's share in the divine mind but it can be used to interfere with or pervert *Natura*. In establishing man's place in Nature, this chapter will deal with these problems and justify us in speaking of Pliny's overall positive and optimistic attitude. The key concepts will be examined: what sort of progress does Pliny actually see in man's activities? What do the terms used to describe the factors instigating man's activities—*ratio*, *usus*, and *casus*—refer to? We shall then deal with the negative aspects: man's physical and

[2] Plants are a trivial but healthful subject (*HN* 20. 2). Bks. 20–7 deal with the use of plants in medicine (see Ch. VI *passim*) and they also form a main part of the subject-matter of bks. 12–19.

[3] e.g. *utilitas vitae* (28. 2); plants *vitae salutares* (25. 25). On the general importance of *iuvare mortalem* as a theme in the *HN* see now Citroni Marchetti (n. 1).

emotional frailty which makes life in Nature hard, and his *luxuria* which opposes Nature. Finally, we shall reaffirm the optimistic outlook: Pliny's references to gardens build up a picture of man and nature in harmony.

1. *Man's Progress: Pliny's View*

It has been said that there is an ambiguity in the very term 'progress'.[4] It may refer to technical, or intellectual, or moral advance and an idea of full-scale progress in every sense is rarely if ever to be found in ancient thought. For a number of writers,[5] technical advance is accompanied by moral decline and a suspicion that intellectual and technical progress actually causes the moral decline may lead to opposition to the progress made.[6] Even in a purely scientific context, there may sometimes be doubt as to the value of the new as opposed to the old and tried.[7] Most assertions of progress are to be found in scientific works—the Greek medical writers from the fifth century BC onwards and Roman writers such as Vitruvius—where the context is amoral. Philosophical works contain fewest; a negative attitude is particularly apparent in the philosophies of Pliny's time, which became to varying degrees more centred on moral issues. The Cynics opposed culture as corruption, Epicurus saw his philosophical life of peace as eventually freeing itself from liberal arts and political life, and Zeno thought the arts had no intrinsic value.[8] Seneca's emphasis on the philosophical life as opposed to material culture varies, but at its most extreme, in *Ep*. 90. 24, it classes all the technical and cultural arts as irrelevancies: they are

[4] E. R. Dodds, *The Ancient Concept of Progress* (Oxford, 1973), 2.

[5] '. . . most acutely in Plato, Posidonius, Lucretius, and Seneca' (ibid. 24). In the present context, Lucretius and Seneca are the most relevant. Lucretius sometimes sees the skills of civilization as advancing (5. 332–7), but elsewhere sees invention as creating necessity (5. 1412–15). Seneca can see and approve advances in pure science (*NQ* 6. 4. 2; 7. 25) but applied science is harmful (*Ep*. 90. 7–16, immoral advances in building and weaving techniques). He criticizes any discovery made by a mind whose gaze is directed towards the earth (*Ep*. 90. 13), an implicit contrast to the celestial observations of the philosopher.

[6] Ar. *Clouds* 1420 ff. See L. Edelstein, *The Idea of Progress in Classical Antiquity* (Baltimore, 1967), 36–7.

[7] Edelstein, ibid. 38 mentions dissent on the relative values of old and new remedies in the Hippocratic treatises *On Ancient Medicine* and *On Hebdomads*.

[8] See Edelstein, ibid. 133–46. On Zeno, D.L. 7. 32.

due to *ratio* but not to *recta ratio*. His optimistic predictions of progress occur only in the context of the celestial observations and natural studies, which he considered increased awe and knowledge of the divine.

If the concept of progress generally is a complex one, then Pliny's own attitudes are even more so. We saw in Ch. I. 4 that his awe of the divinity of nature and sense of the inadequacies of the human intellect caused him to impose limits on the one type of progress—celestial investigations—which Seneca positively encourages. Yet this does not prevent him from voicing a hope that the generations are constantly progressing in the context of explanation of the movements of the planets (2. 62). No doubt this is partly because of his humanitarian views on the fear arising from ignorance or false ideas on such topics (sect. 3), but it also hints at an enthusiasm for intellectual theories and the advancement of study for its own sake. He does not always take the view that there has been a continual line of decline in morality either; there are a number of remarks which compare the present day favourably with those of the earlier Principate or late Republic (sect. 4).

However, the passage in 2. 62 just cited remains isolated: the two passages which discuss the concept of progress at any length, *HN* 2. 117 ff. and 14. 1 ff., take a form similar to a number of other first-century discussions of the decline of literature and learning.[9] Peace is seen as creating a state of intellectual lethargy, to which may be added the *desidia* produced by the spread of luxury corrupting the *artes*. 'Men's morals have aged' (2. 118) says Pliny, and 'pleasure has begun to live while life itself has ceased to be' (14. 6).[10] Yet, if we must set this picture of moral and intellectual decline against the more positive attitudes mentioned in the last paragraph, there remains the fact that Pliny's view of life gives more opportunities for progress of one sort or another than Seneca's more strictly philosophical understanding of the concept. Pliny equates *vita* with an activity which is not intellectual in a narrow and exclusive sense, like Seneca's in *Ep*. 90. *Artes* are linked in a general way with the use of

[9] See D. A. Russell (ed.), *'Longinus' On the Sublime* (Oxford, 1964), 185–6 on such decline as a *topos* of this era.

[10] For *luxuria* as a corruption of the activity inherent in Pliny's concept of *vita*, see sect. 4.

intelligence but some are more worth while than others: '[Now] the minds of men are occupied with other matters and the only *artes* cultivated are those of avarice. Previously the sovereignty of the nations was self-contained and so their talents were similarly constricted. Thus a certain barrenness of fortune made it necessary to make use of the gifts of the mind and countless kings received the homage of the arts . . .' (14. 4). The researches which have suffered include those connected with the practical science of agriculture (14. 3): '. . . so much more fruitful was the care bestowed by our ancestors, or so much more successful was their diligence, when a thousand years ago at the dawn of literature Hesiod began to set forth rules for farmers in which researches he was followed by a number of writers . . .'. Pliny is, however, willing to praise the *pax Romana* elsewhere because it allows the transportation of the utilitarian commodities vital to 'life' (27. 3) as well as the non-essential luxuries, which can have a debilitating effect.[11]

In *HN* 29. 140, Pliny cites as an excuse for including some disgusting remedies made from beetles 'admiratio naturae *priscorumque curae* . . .'. This tendency to elevate the past at the expense of the present makes Pliny's attitude appear more negative than it really is. For instance, he makes general points about factors which discourage the dissemination of knowledge but suggests they are recent phenomena, whereas we should suspect that they are perennial. One such factor is that men do not pass on their learning because sole possession of it gives them more *auctoritas*.[12] Another important one is the complaint that much of the subject-matter of the *HN* is regarded as being too base and trivial for serious study. Such painstaking research into unromantic and unelevated facts (*rerum humilitas*, 14. 7) does not bring popularity (*gratia*, 29. 1–2).

The contrast is well expressed in *HN*, pref. 16, where he praises those 'who have preferred the usefulness of giving help to the popularity of giving pleasure'. In the first of the 'decline' passages, he suggests that only men of earlier times were content with the sole reward of helping others (2. 118). However, the same passage conceals what may be termed Pliny's progressive

[11] Below, sect. 4, Chs. V. 2. D, VI. 1. B.

[12] *HN* 25. 1–2, 16, see Chs. III. 2. B. iii, VI. 2. C.

ideal. Today, he says, no one adds to knowledge by means of original research and even the discoveries of our predecessors are not being studied; those men for the most part had no reward other than knowing that they would benefit future generations. Ideally, then, each generation should assimilate the legacy of earlier knowledge and add new material before handing it down. Moreover, other passages in the *HN* suggest it is more than an ideal. Pliny sees himself as adding recent knowledge to the inheritance. Often, this knowledge is not entirely original but is indicative of a desire to update old ideas, as in *HN* 17. 137, where he mentions a new method of grafting devised by Columella, of which he has not yet acquired a complete account. Recent discoveries in the field of herbal remedies are often recorded (21. 48, 25. 37, 25. 85, 26. 38). With an account of planetary movements, however, he claims originality as well (2. 71).[13]

Does Pliny believe that the incentive to acquire new knowledge is anything more than humanitarian or utilitarian? The ideal outlined in *HN*, pref. 16, is that of overcoming difficulties for the purpose of helping others, a literary and intellectual corollary of the practical Stoic ideal of Hercules. *Cura* and *diligentia* in the acquisition and transmission of knowledge are frequently admired in the *HN* (e.g. 2. 43; 3. 17; 7. 8) often with reference to men of earlier times (27. 4) but also with reference to his own research (6. 23). Yet it is difficult to exclude an element of intellectual curiosity, an appreciation of the knowledge and the process of acquiring it for its own sake. In *HN* 2. 247, Eratosthenes' measurement of the earth's circumference is said to be overambitious (*improbum*), but 'achieved by such subtle reasoning (*subtili argumentatione*) that one is ashamed not to believe it'. The detailed account of the mathematician Timaeus' theory of the rising of the Nile suggests that Pliny admired his 'occultam rationem' (5. 55) although he makes an objection at the end of the account. Pythagoras' calculation of the distance between sun and moon is that of a man 'sagacis animi'. Pliny contrasts it

[13] He does, of course, claim innovation in a traditional literary manner, whilst being aware of the tradition behind him. His accumulation of past knowledge is the greatest yet (pref. 17) and he also refers to an almost untrodden path before him (pref. 14). In pref. 20, 'I do a favour to future generations who I know will take issue with me, just as we have done with our predecessors', we have an example of his awareness of reaction and innovation and not just the passive acceptance of earlier knowledge.

with his 'frivolous' theory that the planets are in some form of harmonic ratio (2. 83–4).

Are these cases exceptional? Pliny does not always use *subtilitas* in a very favourable sense (e.g. 18. 205, see below, sect. 2). The Latin word *curiositas* was ambiguous, very often implying study which went beyond the limits of useful necessity, whether moral, intellectual, or social.[14] Pliny was, as we have seen, aware of the ambiguities of Eratosthenes' study (Ch. I. 4). He feels it necessary to condemn ambitious investigations which lack a firm basis of evidence. These are suspect, not only intellectually, but also morally, since they could decrease respect for the greatness of *divina Natura*. They are also likely to be more pretentious than useful. Knowledge of general principles may be quite sufficient for the useful purpose of banishing ignorant fear. Further curiosity is not only *improbus* but also a waste of time, 'paene dementis otii est' (2. 85). In *De Tranq. Animi* 3. 6–8, Seneca closely links humanitarian usefulness with intellectual responsibility and reverence towards the natural world. Quoting the first-century Stoic philosopher Athenodorus, a friend of Cicero and teacher of Augustus, he draws a contrast between studies conducted with a view to the needs of human society—'nec tibi gravis eris nec *aliis supervacuus*'—and those conducted in complete isolation from other men. In the latter situation, a man will experience *inopia rerum agendarum* and will start to spend his time badly. The examples given include 'thrusting back the sea' and 'causing water to flow despite the obstacles of nature'. Such artificial elaborations devised by the rich for their private estates were criticized by Horace (*Od.* 3. 1) and would be classed by Pliny as perversions of *Natura*.[15]

Pliny was well aware that his own collection of facts in the *HN* was open to the criticism of intellectual superfluity.[16] We see this in his literary defence of his work.[17] He can justify himself in two ways, one specifically stated and the other implied generally.

[14] See further A. Labhardt, 'Curiositas', *Mus. Helv.* 17 (1960), 206–26.

[15] See further sect. 5, and contrast Pliny's enthusiasm for utilitarian engineering (36. 121 ff., Ch. V. n. 59).

[16] See Sen. *Brev. Vit.* 13 on trivialities in historical writings. For other examples, see Labhardt (n. 14).

[17] See Ch. V. 1. B.ii, also the preface to the *HN* and e.g. *HN* 11. 4 and 29. 28 (Homer did not disdain to talk about the fly 'inter proelia deorum') on trivial subject-matter.

He often answers the criticism of intellectual triviality by saying that his 'humble' facts contribute to a better understanding of the greater whole, *Natura*, either by the combined weight of such details or by the function of certain individual details as microcosms of the whole.[18] The second justification, implied throughout the *HN*, is that his concept of what is useful is extremely wide in scope. It can include everything, from the naming of countless species of trees and plants, with its practical implications for agriculture and medicine, to those theories of celestial phenomena which are permissible because they are plausible and remove unnecessary fear. Some things appear useless, but it is only a matter of time before their uses will be discovered (25. 15, see Ch. I. 3).

Overall, then, Pliny does not approve a curiosity which goes beyond certain limits. However, his criterion of usefulness is a wide one, as is shown by the *HN* itself. The range of factual material just described is heralded in the Preface by the ambitious claim that the author will be the first to treat in one work all the subjects dealt with separately by his predecessors (§ 14). Intellectually, man must not overreach himself. Yet, as long as he does not overestimate his own mental power (*ratio*) and in the process underestimate *divina Natura*, it is possible to view *curiositas* as a positive, even a vital, asset. Indeed, Seneca (*De Otio* 5. 2–3) says that *Natura* gave man his *curiositas* precisely in order to provide herself with an appreciative spectator. The wealth of material offered by his ambitious and wide-ranging subject left Pliny with neither time nor inclination to toy with recondite theories. There was plenty to provide for the spectator of *Natura* without them. The result is a work of progress in the broadest sense.[19] Pliny regards the *HN* as a step forward—in making more complete man's appreciation of *Natura*: 'Hail

[18] See Ch. I. 3 and nn. for microcosm and references; *HN* 11. 4 for individual insects as microcosms of nature; *HN* 7. 7 on individual elements contributing to the greater completeness of the whole.

[19] For a different emphasis, see Th. Köves-Zulauf, *Reden und Schweigen: Römische Religion bei Plinius Maior* (Munich, 1972), 315 ff. He sees, in Pliny's *curiositas*, respect for *cura* and *diligentia*, and for the knowledge of his predecessors, a nostalgia for a small world, a *horror infiniti* (333). It is undeniable that there is a desire for completeness and certainty in the very scope and concept of the *HN*: Pliny's concern to eradicate human fears by replacing ignorance with factual certainty (see next section) does, on occasion, have a rather neurotic, morbid undertone, as in certain aspects of his interest in medicine (see Ch.

nature, mother of all creation, and mindful that I alone of the men of Rome have praised thee in all thy manifestations, be gracious towards me' (37. 205).

2. *Human Ability: The Factors Instigating Man's Activity*

It is helpful to define man's activity more specifically by examination of the terms used to describe the factors involved. One modern authority states that, in Pliny, man is helped by Nature, by chance, *casus*, and above all by experience, *usus*.[20] Another believes that Pliny rejects the theoretical thinking implied by *ratio* in favour of practical learning—*usus* and *observatio*.[21] The former definition is incomplete and over-schematized, while the evidence for the latter view does not stand up to closer examination. If the terms *ratio*, *usus*, etc. are to be useful, we need to know what Pliny means by them, both in themselves and in relation to one another.

Of all the terms, *casus* is the most difficult to define. In one passage, Pliny appears to separate it from Nature—*HN* 17. 101 'magister alius et paene numerosior'. In another (27. 8), the two seem more or less identical: 'This chance, therefore, is that god who has made most of life's discoveries, this is the name of the being by whom is meant the one who is at once mother and mistress of all creation'.[22] In this passage, the distinction between *casus* and *Natura* is less important than that between *casus* and the pair *ratio* and *usus*, as is made clear by the previous line. 'Who doubts that this remedy was discovered by chance . . . since wild animals are unable to communicate theoretical and practical knowledge among themselves?' Several other passages suggest *casus* is closely connected rather than contrasted with nature: '. . . reliqua genera casus ingenio suo excogitavit . . . ac docuit'

VI. 2. E). But his concept of *divina Natura* (Ch. I. 4) and the limits of human intelligence suggests there remains an area of uncertainty.

[20] I. Lana, 'Scienza e politica in età imperiale romana', in *Tecnologia*, 21–43.

[21] M. Frederiksen, 'Plinio il Vecchio e l'agricoltura in età imperiale romana: Gli aspetti tecnici ed economici', in *Tecnologia*, 81–97.

[22] Even if 'hoc habet . . . magistra' is a later addition (see e.g. W. H. S. Jones in the Loeb edn. of the *HN*, ix, 392), *deus* makes the same point since we have seen that the only candidate for description as either *magister* or *deus* in Pliny is *Natura*. On this passage, see further Ch. IV. 2. A.

(17.123) uses phraseology more appropriate to a rational mind than to a random factor. In that case, since *casus* is definitely distinguished from the attributes of man, *ratio* and *usus*, in *HN* 27.8, it remains that it can only be connected with the other rational power, *Natura*.[23] Perhaps *casus* is best defined as *Natura* working on some accidental circumstance of human or even animal action. *HN* 17.101 concerns the role of *casus* in the initial discovery of grafting. The farmer who puts ivy-wood bases on his fence-posts does so with the intention of preventing rot, not of creating a graft. The only difference between this and the direct action of *Natura* in the process which taught *inoculatio* in *HN* 17.99 is that here the bird which deposits the seed enacts a passive role in a wholly natural cycle.

Finally, we should note two intriguing passages. In *HN* 25.16–17, Pliny introduces a remedy for hydrophobia with these words: 'We should add that there is no sure means of discovery (*ratio inventionis anceps*), for even of those already known, in some cases chance has been the finder, but in others, to speak the truth, it was a god.' The admittedly cautious reference to 'god' is none the less unusual in the sceptical Pliny. Could this be *Natura* working directly in her divine capacity and not in response to outside activity? Certainly, a more general statement in *HN* 27.1–2 identifies 'the gods' with *divina Natura*. Admiring the plant remedies recorded by his predecessors, Pliny says that 'even the bountiful generosity of Nature might seem to have been surpassed by these men if the discoveries had been the result of human endeavour. But in fact it is clear that this generosity is that of the gods (*deorum fuisse*), or at least divine (*divinam*), even when a man was the actual discoverer and that the same mother of all things both produced the herbs and made them known to us.'

It is interesting that Pliny should sometimes slip into more traditionally religious vocabulary when discussing the obscurer side of Nature's power, as in these two passages. On one other occasion, he even goes so far as to dismiss any attempt at rational

[23] It is this inherent rationality which distinguishes Pliny's Chance and its relation to his *Natura* from the chance factor in the haphazard workings of Epicurus' irrational and impersonal *Physis* as described in *Ep*. 1.75: C. Bailey, *Epicurus: The Extant Remains* (Oxford, 1926), 46–7. Here, the random functioning of Nature is then worked upon by the reasoning of man.

explanation. 'By whom was it discovered and by what chance was it revealed?' he asks in astonishment of the idea that goat's blood cracks diamond (37. 60). 'It is to divinities (*numina*) that we must attribute such inventions, nor must we seek *ratio* anywhere in Nature, but will (*voluntas*).' Pliny's Latin is here obscure in its brevity but he appears to find his usual explanations inadequate for this bizarre phenomenon. *Casus* is passed over. 'Not seeking reason' anywhere in nature rules out use of man's *ratio*. It would also make attribution to the higher *ratio* of *Natura* rather paradoxical, for Pliny wishes to bring out the obscurity rather than the ultimate 'reason' of Nature here. The passage is exceptional but serves as a useful reminder that the sceptical and rational Pliny is not always as far away from popular religious ideas as he sometimes claims (Chs. III. 1, VI. 2. C, cf. III. 2. A. iv).

In general, then, Pliny favours the view that *Natura* is somehow responsible for most of life's discoveries. This raises a problem; if *Natura* is so often their instigator, how original are man's activities? *Natura* is cited as man's teacher in *HN* 17. 58 and 99 among other passages, which suggests that man learns and imitates without real invention.[24] Does the idea of *Natura* as the great *artifex* lessen Pliny's inclination to examine the technical steps when discussing certain processes or inventions? The concept in itself should not; the Stoic Posidonius showed interest in techniques (Sen. *Ep*. 90). A few, agricultural, inventions are mentioned in the *HN* and this at least marks Pliny out as more interested in such things than his contemporary agricultural writer Columella (Ch. V. 1. A. iii). The technical descriptions are not extensive but then neither would we expect them to be in a general encyclopaedic work. Pliny is concerned with why rather than how they have evolved, with their effect on man's life in nature. The harvester and ploughshares in *HN* 18 are mentioned because they save man labour. Technical detail is given only in so far as it helps to explain this purpose. Finally, the passage in which Pliny lays most stress on Nature not man as the true inventor (27. 1–2) refers to the initial creation of herbals and

[24] J. M. André, 'Nature et culture chez Pline', in id. (ed.), *Recherches sur les artes à Rome* (Publ. Univ. de Dijon, 53; Paris, 1978), 7–17, quotes Cic. *Leg*. I. 26, 'artes vero innumerabiles repertae sunt *docente natura*'. Note that the passage continues with a reference to the part played by man's *ratio*: '. . . quam imitata ratio res ad vitam necessarias sollerter consecuta est'.

the manifestation of their basic properties: 'et genuisse haec et ostendisse [*Naturam*].' To get a finished ploughshare from the initial gifts of Nature, wood and iron, requires rather more original thought from man. Pliny, who lays such stress on the development for good or evil by man of *Natura*'s gifts (Ch. I. 3 on iron), surely recognizes this.

What value, then, can be attached to *ratio*, the power of theoretical thought?[25] If man's capability of originality is in question, is there some justification in suggesting that Pliny values *usus*, practical learning, above *ratio*? On the second point it is certainly true that, in *HN* 26. 11 *usus*, as practical science, is contrasted favourably with theoretical learning in schools: 'uso efficacissimo rerum omnium magistro'. There is a similar upholding of *observatio* in *HN* 18. 284: ancient man's *observatio* was no less ingenious than *ratio* is now. Other passages, however, suggest that they are complementary rather than opposed. In *HN* 18. 290, they form a partnership for complete knowledge: 'It is enough for me to have expounded the theory (*ratio*); the rest will be ascertained by the personal observation (*observatio*) of the individual.' *Ratio* indicates the general theory, *observatio* the filling-in of particularities.[26] Seneca had said that *ratio* was the understanding of principle (*Ep*. 121. 10–11). Moreover, passages which have been cited as condemnations of *ratio*[27] are not in fact

[25] See *OLD* s.v. *ratio*. There are a number of different senses, most of which may be located in the *HN*. The first listed by *OLD* is 'reckoning' or 'calculation'. All, however, may be said to share the basic ideas of 'logical principle' or 'systematic process'. Thus, *ratio* is the faculty of reason itself, e.g. Cic. *Luc*. 124, 'trisne partes habeat (animus) . . . rationalis, irae, cupiditatis', or the act of using this faculty, 'omnis opinio ratio est, et quidem bona ratio si vera, mala autem si falsa est opinio' (Cic. *ND* 3. 71). Pliny is using it in this sense when he contrasts it with *usus*, as discussed in the text; cf. Cic. *Amic*. 52, 'de amicitia, quam *nec usu nec ratione* habent cognitam, disputabunt'. More widely, any systematic exposition may be termed *ratio* and Pliny uses it of a descriptive account, '[lac] bubulum biberent pthisici . . . diximus in ratione herbarum' (28. 125). It can also refer to any guiding principle or system, whether of the construction or workings of a particular object (7. 213, 'facti horologi rationem'); a method of doing a particular job ('arandi rationem' (18. 182)); or the guiding principle or law of nature herself ('quaecumque est ratio vitalis, illa inest . . . toto in corpore', 11. 10).

[26] It seems clear that *caeca subtilitas* in *HN* 18. 205 ff. is not a criticism of theory *per se*, but of theory which is not in partnership with the practical *observatio* which can allow for consideration of particularities and exceptions to the theoretical rule. See Ch. V. 1. B. i.

[27] By Frederiksen (n. 21).

such if viewed in their wider context. Of those singled out, *HN* 3.122—'pudet a Graecis Italiae rationem mutuari'—is really an anti-Greek remark directed at the necessity of borrowing a Greek account of Italy. Another, *HN* 11.8—'it is my purpose to point out the manifest properties of things, not to search for doubtful causes'—occurs in the context of animal properties and is probably no more than an announcement that Pliny is not going to base his discussion of animals on the specific Aristotelian theory of causes.[28] Finally, *HN* 17.9—'It is not my purpose to give great attention to superfluities'—needs to be set in context. It comes at the end of the following passage: 'Now I shall set forth the right and perfect method (*rationem*) of cultivation and for that reason I shall ignore what is common knowledge and what I observe to be established, but I shall concentrate on uncertain and doubtful matters where practical life experiences problems.' Rather than condemning *ratio*, Pliny is doing the opposite. All he is leaving out as *supervacua* are the already established points of theory. He shows his approval of a solid and complete theoretical basis by his expressed intention to discuss the dubious and difficult points and by implication place them on a firmer foundation.

As to the question of originality, there is no escaping the ultimate ambiguity of *ratio*. As part of the divine *ratio*, that is, as part of *Natura* herself, man's *ratio* would seem incapable of true originality. His behaviour must always be in some sense a reaction to *Natura*. Any originality consists of the ability to choose to pervert rather than to follow her. Apparent criticisms of *ratio* by Pliny are in fact directed at this misuse. The kinship of man's *ratio* with the divine does not imply even a theoretical equality either. Man is mortal and full of imperfections (7.1 ff.). No other animal has *vita fragilior* (7.5, below, sect. 3). Ironically, the only animal to possess true intelligence is the one most subject to emotional and psychological disturbances (7.5). Ultimately, man's intellectual activity is subject to his physical imperfections.

Two passages serve to illustrate these conclusions. First, *HN* 22.117–18, which deals with the compound medicines devised by man from natural materials. 'Nature produces works that are

[28] Cf. Arist. *PA* 1.5, 645^{a}25 ff. See A. Ernout in his commentary on *HN* 11 (Budé edn.; Paris, 1947), 17–18.

absolutely perfect: a few ingredients are chosen with purpose not by guesswork (*ex coniectura*), so that dry substances may be modified by some fluid to help their passage . . . but to collect and mix the active ingredients with minute precision (*scrupulatim*) is not human guesswork but human impudence'. It is not clear whether *Natura* or man makes the approved simple combinations, but it seems logical to take the contrast as being that of the right and the wrong ways for man to combine the natural elements. *Natura*'s ingredients are *perfecta*, so need little or no modification. What modification there is should take the form of improving texture and consistency, not of improving the efficacy, which is already perfect. *Natura* alone knows for certain the qualities of her ingredients: 'ex causa non ex coniectura' is to be contrasted with 'coniecturae humanae opus'. But this is not merely condemnation of *coniectura*. By weighing out the ingredients *scrupulatim* man has ignored the limits of human *ratio* and has given his guesses a false certainty. Perverted human *ratio* challenges perfect *naturae ratio*, and this is *impudentia*.[29]

The passage on the intricacies of weighing out the fifty-four ingredients prescribed for the Mithridatic antidote (29. 24–6) also shows that such exactness is really a sign of man's inherent mortal imperfection. Pliny believes that the choice of ingredients and their measurement is a result of false cleverness, whose end product could not have been produced by rational thought. 'By which god, in heaven's name, were these proportions demonstrated? Such subtlety was not within human capabilities, but is a blatantly showy display of art and a monstrous gimmick of science.' It is close to magic, the ultimate irrationality, which is also the most culpable and deliberate attempt to interfere with and pervert the workings of the divine (see Ch. III. 2. A. ii). *Ratio* is not in itself disapproved of, despite the fetters imposed by the Plinian criteria of practical usefulness and humanitarian motives. Man can, however, pervert *ratio* itself by reason of his mortal imperfections.

[29] Pliny is particularly opposed to ingenious theory in medicine because he is biased against Greek cleverness generally and has a layman's suspicions of doctors (Ch. VI). On this passage, see esp. Ch. VI. 2. B; also Chs. I. 4, III. 2. A.i–ii, V. 2. D, VI. 1. B.

3. *Human Frailty: Pliny's Humanitarianism*

Pliny's awareness of human frailty and the paradox of such weakness existing in the supreme creation is most obviously illustrated in the opening paragraphs of *HN* 7, which is devoted to a description of man. The passage belongs to a tradition of writing on the differences between man and animals,[30] but it is important to note that its purpose is not to exalt animals at the expense of man. Pliny is concerned with the paradox contained within man: human strengths and weaknesses are equally obvious and are stressed by rhetorical antitheses.[31] The key question of the passage, whether *Natura* is a mother or a stepmother to man, is left open.

How, in Pliny's view, is man's life *fragilis* (7. 5)? In what ways can man overcome this and remain the greatest creature in nature? Seneca, in a similar passage on man's weaknesses, also emphasizes the word *fragilis*: 'Quid est homo? vas . . . fragile . . . Quid est homo? inbecillum corpus et fragile . . .' (*Ad Marc.* 11. 3). Certainly, Pliny is well aware of physical frailty: later in book 7 there are hints that the most brilliant intellect when encased in a feeble body is not to be taken too much for granted. Physical injury can cause loss of memory, which is 'necessarium maxime vitae bonum' (7. 88) but 'nec aliud est aeque fragile in

[30] Like Cicero (*Rep.* 3. 1. 1 = Aug. *Contra Iulianum Pelag.* 4. 12. 60), from whom he got the striking *mater–noverca* antithesis (cf. Hesiod, *Op.* 825), Pliny is concentrating on the paradox within man himself—physical weakness combined with mental power. The growing trend in philosophy towards moral questions helped to sharpen the man–animal differentiation in man's favour, on the principle of the mind's superiority to the body. Philo also uses a stepmother simile but doubts whether physical superiority is a real advantage (*Post. Caini* 160–2). Seneca discussed the moral superiority of man, the sole possessor of reason in *Ep.* 124, introducing his theory thus: 'quaeritur, utrum sensu comprendatur an intellectu bonum? Huic adiunctum est in mutis animalibus et infantibus non esse.' Plutarch's eclectic generalizations stress the beasts' physical superiority but man's mind makes him dominant (*De Fortuna* 98C–E). For earlier ideas, see Plato, *Protag.* 321C ff.; man's mental power is actually superior to the physical strength of the beasts. The possession of hands is the consequence and not the cause of superior intelligence as Anaxagoras had argued (Arist. *PA* 10, 687^a7–10.). See further Ch. IV. 2. A.

[31] Note in particular the chiastic word-pattern in '(natura magna) saeva mercede contra sua munera, ut non sit satis aestimare, parens melior homini an tristior noverca fuerit'. In 'itaque feliciter natus . . . flens, animal ceteris imperaturum', the paradox is emphasized by alliteration.

homine' (7. 90). In old age, not only do the senses of the physical body die, but so does *sapientia* (7. 168–9)—good reason to call death a benevolent gift of *Natura* according to Pliny. Yet even here, where animals would seem better off than man, he claims that Nature can still single out man for a unique degree of physical immunity from lightning in order to make up for this very weakness (2. 145).

The really important human frailty for Pliny is that of the emotions. The mind is capable of moral choice; *ratio* risks being perverted by the emotions. Pliny's moral outlook is traditionally Stoic. The good things of life, wealth and power, are to be preferred but not sought after to an 'unreasonable' extent. Pliny scorns the common excuse that man cannot be wise (*sapere*) all the time; he is far too prone to misuse his mind in self-deception ('vana mortalitas et ad circumscribendum se ipsam ingeniosa . . .', 7. 131). Nature is not criticized for making man as he is. Good and bad things come to him largely as a result of outside circumstances, the changes of fortune. This stress on Fortune (Ch. III. 2. c) rather than Nature accords with the Stoic view, which took man's physical weaknesses for granted and considered that they could be overcome if the innately superior mind set itself on philosophical goals.

Pliny would have agreed with Cicero (*De Fin*. 2. 112) that for man to use his superior gifts to attain mere pleasure would be like Xerxes marshalling all his forces just to obtain Attic honey. However, his practical outlook would allow of more activity than contemplation of the universe. Admiration for all manner of achievements is expressed in book 7 (Ch. I. 2 and above, sect. 1). Pliny's realistic outlook is also reflected in his moral judgements. In real life, the greatest men will combine the faults of power with the greatest morality, as well as sheer ability, of mind. His description of Caesar's career, for instance, covers both his triumphs for the state and the evils of power and wealth (7. 92–3). Yet it also allows for the comment that his burning of Pompey's papers after Pharsalus was 'vera et incomparabilis invicti animi sublimitas' (7. 93). The idea of the use and abuse of *ratio* underlies these balanced assessments of the power of genius for good or evil. The realism of Pliny also gives rise to a sense of proportion in his criticism of human imperfection. The deprecation of pride (7. 3) in such a weak creature as man is not, perhaps, as literally universal as it might have been in a more strictly philosophical thinker. Certainly, when the *animalium superbissimum* is again reminded of the frailty of human existence (7. 43), it is clear that Pliny refers not to the ordinary thoughtless man, but to the *tyrannus* and *carnifex*: 'One feels pity and shame when reckoning how trivial is the origin of the proudest of animals. It is from these beginnings that tyrants are born and the mind that contrives slaughter.' His object of attack is the extreme case, where misuse of the mind leads to the evil of abusing the law of humanity.

Pliny, then, sees man's emotions as his true weakness without being narrowly condemnatory. Indeed, his description of such emotions is tinged with sympathy. Towards the end of the introduction to book 7, he moves from the physical to the emotional weaknesses: man alone has *luctus*, *luxuria*, *ambitio*, *avaritia*, *immensa vivendi cupido*, *superstitio*, and *sepulturae cura atque etiam post se de futuro* (7. 5). He sums up: '. . . no creature has a more vulnerable life, none has stronger desires, a more confused fear, a more ferocious rage'. Perhaps Pliny had in mind the Stoic theory of the four emotions—pain, fear, desire, and pleasure (D.L. 7. 110). What is interesting here is Pliny's choice of adjectives. They suggest pity for the contradictory elements in man, the havoc which *pavor confusus*, for instance, or *rabies acris* can play with *vita fragilis*. Seneca thought that such emotions should be rooted out at all costs but Pliny is closer to the compassion expressed in the words of a later Stoic, Marcus Aurelius: not only is that which proceeds from men worthy of love because they are akin, but sometimes also of compassion because they are ignorant of good and evil (*Med.* 2. 13). There was in later Stoicism a growing tendency to lay less emphasis on the strict philosophical relationship of the emotions to body and mind and more on the actuality of the whole man and the whole of mankind. The rational creature is a civic creature (*Med.* 10. 2); man's chief characteristic is fellowship (*Med.* 7. 55) an idea already found in Cicero and Seneca and often attributed to Posidonius. Various sayings of Epictetus also illustrate this compassionate trend (e.g. 3. 13. 3–4), and, despite his continual exhortations to live a philosophical life by freeing oneself from overdependence on the sense impressions, his approach is often general and not limited to a loftiness attainable only by a few.

A passage in Seneca (*Ad Marc.* 7.4) suggests that fear is the basic emotional disturbance: the emotions arising from poverty, grief, ambition, etc., may all be summed up as fear of things which are not to be feared. Freedom from fear appears in Epictetus' list of the qualities of freedom[32] and is a common philosophical theme, especially in the form of the ultimate and universal fear of death. In a rhetorical paradox (7.190) Pliny treats man's *cura de futuro* (7.5) from the more positive view of a craving for life rather than a fear of punishment which Lucretius had tried to combat.[33] In spite of his rational scepticism, he displays an appreciation of the attractions such a belief has for the mass of mankind, rather than an unbending Stoic reserve or Epicurean detachment. Although he offers the opinion that life after death will bring the same troubles as life before death, and will allow the soul no repose, his language displays sympathy with a 'mortality greedy for unending life', that 'pleasant but fanciful idea' (*dulcedo credulitasque*).[34] Pliny sees the problem in terms of a misguided *avaritia* rather than a misguided *pavor*, which is surely in keeping with his positive attitude to life generally.

We need to look elsewhere in the *HN* to see Pliny concentrating on the problem of fear and how to allay it. Most important is the passage mentioned in Ch. I.4 on the astronomers who release men from irrational or false ideas about the workings of *Natura*: 'Mighty heroes, of loftier status than mortal men, who have discovered the law of these great heavenly bodies and freed the wretched mind of man from fear—all hail to your genius, interpreters of the heavens, discoverers of a theory whereby you have conquered gods and men' (2.54). This passage on the men who have explained eclipses was influenced by Lucretius' praise of Epicurus as the illuminator of the mysteries of nature, thereby freeing men from fear: he raised man from *religio* (*De Rerum Natura* 1.62–79), reasoned with a godlike mind ('divina mente',

[32] Epict. 4.3.7.

[33] Cicero, too, had approached the problem of fear from the positive side, since he considered that concern with the fear of punishment after death was old-fashioned (*ND* i.86). Nevertheless, Pliny echoes Lucretius' language and ideas several times in *HN* 7.188 ff., e.g. death as tranquility (190, *securitas* and *quies*). There are echoes of Lucretius 2.5–6, 3.87–93 in *HN* 7.189, 'puerilium ista deliramentorum . . . commenta sunt'.

[34] Yet he does admit that death is very often a blessing in that it offers an escape from physical hardship. See further Ch. VI.2.E.

3. 15) and is 'rerum inventor' (3. 9). Pliny may also have recalled 1. 70–1 when he wrote of Anaxagoras: 'rerum fores aperuisse' (*HN* 2. 31).[35]

There are important differences between Pliny and Lucretius on investigation of the cosmos. Lucretius regarded the strict limits on the power and extent of nature as being discoverable by human science (1. 77, 5. 90). Pliny, as we have seen, thought that human science was liable to overestimate its capabilities and thereby to underestimate a divine and often obscure *Natura*. He praises such investigation only in so far as it removes fear. Like Lucretius, however, he sees the mistaken fear of natural phenomena as misplaced *religio*, unfounded beliefs in the traditional gods and fatalistic creeds. Concern for man's fear is voiced in a passage dealing with a popular superstition (11. 273): '. . . ne auguria anxie quaerat' (Ch. III. 2. B), and *superstitio* was listed as another of man's emotional weaknesses in *HN* 7. 5.

Both Lucretius and the Stoics saw reasoned argument as the way to overcome the irrational emotions.[36] Seneca said 'ratio terrorem prudentibus excutit' (*NQ* 6. 2. 1). It was a struggle, however, and both Pliny's and Lucretius' verbal imagery suggests this. The success is a 'victory'. For Lucretius, Epicurus is the victor of superstition (1. 63–78). For Pliny, the heroes who interpret the heavens are conquerors: 'vicistis' (2. 54). Seneca (*Brev. Vit.* 14. 2) summed up the aim of Stoic philosophy as *hominis naturam vincere* and Pliny's heroes were *supra mortalium naturam*. The condemnation in both Lucretius and Pliny of the aggression of man towards man adds point to their use of metaphor of strife and victory; true victories are won in attempts to help not injure others.[37] The rational mind is the weapon that

[35] See Bailey on 1. 66–7.

[36] Lucr. 1. 72; 2. 53 ff., Galen, *Plac. Hipp. Plat.* iv. 7 (i. 286 De Lacy).

[37] Lucr. 5. 1289–96, 1305–7, 1347–9 ~999–1000. Pliny's unfavourable contrast of man's attacks on his own kind with the behaviour of the animals is exaggerated; there are examples of such aggression among animals of the same species in 10. 25–6, 10. 198–9. However, man's peculiar tendency towards such aggression may be an instance of the microcosm reflecting what is to be found in the darker side of nature, *Natura saeviens* (see Ch. I. 3). Cf. 32. 2 *rabies mundi*, in a description of a stormy sea (Ch. I. n. 52, cf. Ch. V, intro. and 2. E), with which we may compare the citation of *rabies* as one of man's debilitating emotions in 7. 5. In 7. 18, *Natura* is said to have given certain men the habit of cannibalism and the power of the evil eye 'ne quid usquam mali esset quod in homine non esset'.

wins the victory and repels superstition. Lucretius says that it is roused to action by superstition; stories of threatening gods 'acrem/irritât [*Epicuri*] animi virtutem' (1. 69–70). Phrases such as 'omne immensum peragravit mente animoque' (1.74) and 'ergo vis animi pervicit' (1.72) express the idea of mental exploration and conquest. Pliny hails the *ingenium* of his conquering heroes and their role as *interpretes* and *argumenti repertores*.

Perseverence in the right use of *ratio* produces complete harmony with the great mind of *Natura*. It will be recalled that Pliny praised Caesar for showing 'vera et incomparabilis *invicti* animi *sublimitas*' (*HN* 7. 93) by burning his enemy Pompey's private papers. The minds of the astronomers in *HN* 2. 54 are lofty enough to be 'rerum naturae capaces'. The great Hipparchus was 'consiliorum naturae particeps' (2. 53). This loftiness of mind is the highest proof of man's essential relationship to *Natura*. In the words of Seneca (*Ep*. 104. 23), *Natura* made man *magnanimus*, just as she made some beasts fierce and others crafty; she has given man 'gloriosum et excelsum spiritum . . . simillimum mundo'.

For Pliny, right use of *ratio* conquers the emotions. He uses reasoned arguments against the *avaritia* which makes men hanker after life beyond the grave (7. 188–90). He refers to this desire as *vanitas*, a word frequently used of a major perversion of *ratio*, magic.[38] It is, as was suggested at the beginning of this section, this ambiguity within man which interests Pliny. The physical frailty of man compared to the other animals is a fixed fact, an ambiguity of nature which balances out in practice because of man's *ratio*. If animals have any advantage it is freedom from the struggle caused by the possession of mind and emotion (2. 25): they do not think 'de gloria, de pecunia, de ambitione, superque de morte'. Instead, they think only of *victus*—survival, bare necessity. Man's mental qualities put his *vita* on a different plane: his possession of them means that he has a variety of experience, but his emotions produce distortion. *Gloria* and the rest become exaggerated values obscuring necessity. Money, repute, ambition are all involved in *luxuria*. *Luxuria* is the corruption of the life of vigour but comfort, which encapsulates man's superiority to the rest of the animal world, into the life of enervation and

[38] Magic as *vanitas*: 28. 94, 29. 81, 30. 1. See Ch. III. 2. A. ii.

degeneracy. It is the distortion of *vita* as a whole. Its implications are examined in the next section.

4. Heu prodiga ingenia . . .: *Man's Perversions in Luxury*

The power of man's *ratio* is great and produces a *varietas* which is, again, a trait of *Natura* herself. This *varietas* is a general aspect of man's position as a microcosm of *Natura* (Ch. I. 3), a reflection of her power, and is seen in Pliny's account of the multiplicity of races and the physical differences between them (7. 6–8). More specifically, however, *varietas* is manifested in man's divine spark, his mind.[39] In *HN* 7. 52, Pliny says that the passing thoughts of human parents can produce different features in their children. The *velocitas* of man's mind can produce a multiplicity of images in contrast with the other animals, whose minds are *immobiles et similes*: 'The reason why there are more differences in man than in all the other animals is that his swiftness of thought, quickness of mind, and *ingenii varietas* impress a great diversity of patterns, whereas the minds of the other animals are sluggish and alike for all and sundry in each species.' Cicero, when contrasting the dull uniformity of the beasts, who are born only for survival and procreation, with the higher purpose of man, linked the difference specifically with man's divine kinship ('mortalem deum . . . divinum animal', *Fin*. 2. 40). In *HN* 7. 91, Pliny specifically separates this quickness and brilliance from the moral and philosophical aspects of man's thoughts, and concentrates on the sheer mental power of Caesar. 'Caesar was the most outstanding example of mental vigour. I am not now thinking of valour or resolution, nor of the loftiness which embraces all the heavens but of a native vigour and a quickness winged with a sort of fire.'

Varietas thus has an amoral side, a reflection of the powers of *Natura* in man the microcosm. Yet its connection with mind also suggests a moral ambiguity, man's ability to choose whether to follow or to pervert *Natura*. Plutarch aptly expressed the ambigu-

[39] Variety as opposed to uniformity in remedies is used to show the *ratio* of man in contrast to the mechanical skills and impulses of animals (Origen, *Contra Cels*. 4. 87). Similarly, man's *artes* are contrasted to animal skills (Sen. *Ep*. 121. 23). Cic. *Leg*. 1. 26 talks of *artes innumerabiles*.

ity by the use of the Greek *polutropos* (*De Amore Prolis* 1, 493 D), and it is never far from Pliny's thoughts. It is seen in the language of his exclamation against perversions in luxury: 'Heu *prodiga ingenia, quot modis* auximus pretia rerum' (33. 4). The power and *varietas* of the human mind can have a detrimental effect. It is interesting that a factor in the increase of luxury was multiplicity of resources, 'rerum amplitudo' (14. 4). The inclination to variety leads to the acceptance of non-necessities. Above all, it is a passive acceptance; *pax* has brought *desidia* (14. 3–4). In effect, the power of the mind is not being exercised and, paradoxically, man is reduced by prosperity to a state of bovine passivity rather than divine activity. Learning no longer flourishes; man's mind is as dull and immobile as the beasts'. Thus, *luxuria* becomes a negation of *vita*. At the beginning of the chapter it was stressed that Pliny's idea of life was active. The phrase *opera vitae* is used in *HN* 14. 5 and the comment (14. 4) that 'owing to a certain barrenness of fortune, it was necessary to exercise the gifts of the mind' points to the effort required and to the contrast between the goods of the mind and those of luxury. The change is summed up in *HN* 14. 6: 'voluptas vivere coepit, vita ipsa desiit.'

Luxury is a perversion of the powers of the mind and therefore of the correct way for man to live. All the same, luxury goods are not necessarily bad in themselves, although they can lead to degeneracy,[40] as we shall see in specific cases further on (Chs. V. 2. D, VI. 1. B). For the moment, it is necessary only to remember that Pliny did not condemn culture as a whole.[41] Indeed, the introduction to book 14 stresses that *avaritiae artes* are a perversion, to be distinguished from *artes* in general which are products of the exercising of *animi bona* (14. 4). Secondly, even luxury is tolerable so long as the mind is not deceived by it. However, the tendency is to attach exaggerated and false values to such items: 'quot modis auximus *pretia* rerum'.[42] It is, once

[40] The passivity of the patient is a major reason for Pliny's prejudice against elaborate medicines and health-routines: see Ch. VI. 1. A. See also the contrast in *HN* 14. 5 between the 'liberal' arts and 'servile' attitudes.

[41] André (n. 24), 14 compares this more positive attitude to that in Sen. *Ep.* 114. 55, where man's works are said to be good because based in Nature but are perverted by man's own irrational and inhuman usage.

[42] See also Sen. *Brev. Vit.* 12. 2: Corinthian bronzes which the *furor* of a few makes *pretiosa*. On Pliny's idea that luxury is a distortion of 'life' generally,

more, a case of miscalculation, of measuring things by deceptively reasoned and logical human systems of weights and measures. Pennies, we might say, inflate the value of luxury goods, while inches decrease that of the universe (Ch. I. 4).

It is important to grasp the significance of *pretia*. Pliny's opposition to luxury is not based on purely moral grounds. It is also influenced by a practical concern with material values. His eagerness to give correct monetary prices even for luxury goods such as purple dye has been noted (9. 137).[43] Although he talks slightingly in *HN* 19. 47 of those who grow the plant-dye *rubia* for profit (sect. 5) he does not hesitate to describe how the purple-dye factories could operate more efficiently and therefore, he implies, more profitably—*tempus utilissimum*—if they gathered their shellfish after the dog-days or before the spring, since they would then require less processing to extract the dye (9. 133). Similarly, his criticism of the production of luxury vegetables (sect. 5) does not prevent him from giving a fuller account of their cultivation. Indeed, this is given precisely because it is a means of providing a table luxury, not in spite of this ('quapropter', 19. 139). Clearly, Pliny accepts the manufacture of luxuries in actual life as a means of making a living. Moral outrage is tempered by business considerations.

Side by side with the practical financial criterion is Pliny's fascination with the use and abuse of *ratio*. *Luxuria* often involves a perverted ingenuity on the part of man in rivallng the work of Nature. Thus in *HN* 9. 139, a passage describing the further refinement of the already luxurious purple dye, Pliny brings together man's irresponsible playing with monetary prices and his overingenious elaboration of nature's works: 'Men like . . . to make a game of spending and they like to double their sport by mixing and readulterating Nature's adulterations; for instance, dyeing tortoiseshell, alloying silver with gold to produce electrum-ware, and adding bronze to these to make Corinthian ware. It is not enough to have stolen for a dye the name of amethyst ['sober-stone'], but when it is finished it is made drunk again with Tyrian dye, so as to produce from the combination an outlandish name [sc. *Tyriamethystus*] and a double luxury at

compare Sen. *Ep*. 122. 2, 17–18, on the dissolute revellers who disrupt normal patterns of activity and rest with their banquets, and thus literally 'retro vivunt.'

[43] Citroni Marchetti (n. 1), 142.

once. When they have made shell-dye, they think it an improvement to transform it into Tyrian. Repentance must have discovered this first, the artificer altering a product he disapproved of, but then reason (*ratio*) sprang up and a defect was made into a success by portentous ingenuity and a double path pointed out for luxury . . .' (9. 139–40).

The utilitarian Pliny need not be seen as clashing with a more 'philosophical' Pliny.[44] Pliny's philosophy is practical; it is based on everyday Roman civilized life. He is, in fact, no admirer of primitive simplicity. In *HN* 16. 2, he shows no liking for the *inops ac fera sors* of the Chauci, of which he had, presumably, seen too much in Germany. He thinks it extraordinary that they do not allow the material benefits of Roman rule to outweigh dislike of being subject to another; *libertas* in these circumstances is meaningless.[45] At no time does he suggest that those peoples whom he depicts as following primitive practices today are morally superior to more civilized peoples. Rather, their existence expresses an aspect of the *varietas* of humanity; different peoples are at different levels of progress (7. 206 ff.). Rome is not necessarily corrupted by her civilization; Roman honesty may be admired by foreign peoples (6. 85 ff.).[46] When Pliny describes a Utopia, that of the Hyperboreans, the life extolled is, it is true, rustic, yet it is characterized not by harsh poverty but by comfortable opulence. The inhabitants' choice of *mors beatissima*

[44] *Contra* Citroni Marchetti (n. 1), who sees this as an aspect of his utilitarian viewpoint, which clashes to some extent with his use of the anti-luxury language of philosophy and rhetoric.

[45] Contrast Tac. *Germ*. 35, who ignores the harshness of the lifestyle and praises their moral superiority. They are untroubled by greed or ambition and maintain their greatness not by arms, in which they possess a formidable strength, but by righteous behaviour. Citroni Marchetti (n. 1) contrasts Sen. *De Prov*. 4. 14. On Pliny's 'un-Stoic' stance in this passage, see K. Sallmann, 'Reserved for Eternal Punishment: the Elder Pliny's View of Free Germania (*HN* 16. 1–6), *AJPh* 108 (1987), 108–28, who attributes his hostility to the attempts of a plain Roman soldier to justify Rome's failure to conquer the German tribes in the North.

[46] The only hint of a sophisticated people being contrasted unfavourably with a simpler one is in *HN* 12. 59, where the honesty of the native Arabians who gather the frankincense is set against the fraudulence of those who work it in Alexandria. But Alexandria was noted for general immorality and Pliny attacks its citizens on similar grounds in *HN* 20. 200. In that passage, he quotes a Greek authority on drugs as saying that the only reason that opium does not cause instantaneous blindness is because it is adulterated at Alexandria. Pliny harps on the Greeks' fraudulence generally, especially in their writings (e.g. *HN* 8. 81–2).

(Ch. VI. 2. E) comes after a final burst of luxurious banqueting (4. 89). As for the ascetic life, his attitude is one of baffled fascination rather than approval. He notes as *mira* the rejection of normal human reproduction in the community of the Essenes on the Dead Sea. They are a paradox of Nature, their numbers being supplemented only by other men's *vitae paenitentia* (5. 73). Overall, the characteristics of normal Roman civilized life are seldom challenged. Only once is a 'primitive' system praised to the detriment of a more 'civilized' one. That is bartering, which would dispense with gold, the over-valued material from which is coined the money which encourages the perversion of true values.[47]

These points must be borne in mind when considering luxury in specific fields of human life, as must the paradox of *Natura* herself as *artifex*. One such area is the garden, where man appears to impose his *artes* on the very face of *Natura* and to promote harmony not opposition thereby.

5. *Man and Nature in Harmony: Pliny on Gardens*

It comes as something of a shock to turn from Pliny's strictures on luxury to the extravagant language of *HN* 21. 1, where flowers are produced by a *Natura* 'lascivienti praesertim et in magno gaudio fertilitatis tam variae ludenti'.[48] It is true that he somewhat spoils the effect a few lines later by announcing that flowers, the only non-utilitarian products of *Natura*, are intended by her as a warning to man on the transitory nature of beauty. However, the striking nature of the description is upheld when contrasted with a similar passage in Columella. This, in comparison, is distinctly muted: 'iam versicoloribus anni/fetibus alma parens pingi sua tempora gaudet' (*De Re Rust*. 10. 256–7).

What, then, is the significance of Pliny's words? It seems, first of all, that there is something of a contradiction in praising the abundance of nature in what to modern minds would seem a

[47] *HN* 33. 6, cf. 6. 133, see too 6. 88, Ch. V.2.D. Pliny also mentions, but does not enlarge on, the fact that the Essenes have no money.

[48] An echo of a Stoic idea is present, as often in Pliny. Plut. *De Stoic. Repugn*. 1044 C (*SVF* ii. 1163) records it: 'Nature produces many of the animals on account of their beauty, for she is a lover of the beautiful and rejoices in variety'.

very controlled environment, a very disciplined area of natural growth. If one also considers the fashion for pleasure-gardens in the Late Republic and Early Empire,[49] the flower-garden might rather be expected to incite Pliny to moral criticism. Secondly, we saw in Ch. I. 4 that *Natura ludens* was a being not entirely free of the elements of *Natura saeviens*. Yet the two phrases imply a contrast in mood.

To explain Pliny's explicit approval of an environment where man and nature interact so closely, we need to consider the Roman concept of the garden and the Roman concept of artificiality. What does Pliny criticize in pleasure-gardens and why? What bearing do these views have on the concept of artificiality? We can then envisage the type of garden in which *Natura lasciviens et ludens* and man can be in harmony with each other.

Before embarking on the study of individual passages, a point hinted at in Ch. I. 4 must be remembered. Pliny does not condemn every interference of man with Nature. His attitude depends on whether or not she is perverted or degraded by man's intervention and for what purpose. The generalized statement (*HN* 14. 1) that a large factor in a tree's nature is due to its cultivation acknowledges without approval or disapproval not only cultivation of trees but also that their natures may be changed by different modes of cultivation. This is not, in fact, perversion of Nature to Pliny. Grafting, for instance, although an artificial process, produces many varieties of fruit, one of *Natura's* basic gifts (12. 1 on trees, '. . . hinc primum alimenta . . .'). In one passage it is even treated as complementary to Nature's own work, man being regarded as an appreciative partner and collaborator. There are almost as many trees produced 'cura hominum atque arte' as there are wild ones, 'so bountifully have we repaid our debt of gratitude to nature' (17. 58). The atmosphere is not that of conflict but of friendly competition; indeed, there is a note of reluctance when he says (17. 120) 'nec tamen omnia experimentis adsequi in natura possumus'. *Adsequi*

[49] Pliny would never approve the rejection of the practical in favour of the pleasurable aspects of such places, as is affected by Cicero on one occasion when describing building-progress on a villa of his brother's: 'Caesius thinks you will irrigate thirty-five acres of meadow with ease . . . I'll promise what I better understand, that you will have an exceptionally charming country house . . . (*QF* 3.1.3, tr. Shackleton Bailey)

is revealing, denoting as it does 'catching up with' rather than 'outdoing by opposition'.

Bearing in mind Pliny's ideas on the perversion of Nature and on luxury described in the last section, we may turn to passages mentioning features in ornamental gardens. The plane-tree occurs in a number. It was much favoured in gardens but is criticized by some writers for its sterility; it exists only for the provision of what the moralists considered the luxury of shade.[50] In Pliny, the most forceful condemnation of this tree concerns an evergreen variety (12. 11–12). He regards it as originally an isolated quirk of nature, reproduced by man for the worst motives of luxury—desire of novelty for its own sake. Even more important, however, is his view that, since the plane's greatest commendation is that it gives shade in summer but admits light in winter, this variety's peculiarity is a *vitium*, a fault or perversion. Here, Pliny views shade not as a luxury but as part of a natural cycle.[51] It is undoubtedly true that he also reflects the 'luxury' view, as various tentative statements and apologies show: for example the plane was imported 'umbrae gratia tantum' (12. 6). Less clear in its implication is *HN* 19. 24, where he mentions the use of awnings to shade moss in courtyard gardens. The critical emphasis in fact appears to be on the use of dyed linen—a material which is ambiguous in Pliny. In the present passage, moreover, the linen is treated in a luxurious fashion.[52] That there is, however, a conflict in Pliny's mind between shade as naturally healthful and shade as luxuriously pleasurable is illustrated by a remark just previous to the one on moss. Marcellus placed awnings in the Forum in summer 'ut salubrius litigantes consisterent . . .'. What a change this was, remarks Pliny, from the *mores* of Cato, who believed that the Forum should be paved

[50] Seneca hints at a relaxation from a strict utilitarian view in the pairing of attributes in *Ad Helv*. 9. 1: 'At non est haec terra frugiferarum aut laetarum arborum ferax.' It is, however, part of a hypothetical objection in a philosophical exposition.

[51] *HN*. 17. 91: 'Even this knowledge (of shade) is not to be despised or put into the last category, as shade is either a nurse or a stepmother to a plant': a plant's natural preferences must be fulfilled in this matter if it is to be grown properly.

[52] For linen, see *HN* 19. 3–7 (Ch. V. 2. A). The process of dyeing produces a vivid phrase of condemnation: '. . . in quibus [odoratis floribus] unguento vicisse naturam gaudens luxuria vestibus quoque provocavit eos flores qui colore commendantur' (*HN* 21. 45). Pliny continues the contest imagery in the next paragraph: 'paria nunc conponuntur et natura atque luxuria depugnant'.

with sharp stones. Healthful comfort must not, it seems, be too enjoyable.

On the other hand, Pliny also gives an enthusiastically worded description of the pleasures afforded by a huge hollow plane-tree in Lycia next to a spring and forming by itself a shady grotto: '. . . agros longis obtinens umbris . . . speluncae imagini'—complete with mossy stones. The governor of the province, Licinius Mucianus, had slept inside the tree, 'receiving more pleasure from the pleasant patter of the rain through the leaves than he would have from the gleam of marble, the variety of decorations and the gilt of panelling' (12. 9). This tree would appear to be a phenomenon occurring in the wild—Pliny says it stands by the roadside—so that the comparison is a straightforward one of natural versus man-made beauty. Yet it is not irrelevant to mention it here because Pliny is sufficiently taken by Mucianus' enthusiasm to record it in detail. We could, of course, put this down entirely to Pliny's fascination with marvels of nature. Yet we might surmise that Pliny was also half-consciously thinking of various horticultural devices used by men to imitate the rustic pleasures of a leafy house. He specifically says that this plane tree is like a house: 'itineri adposita *domicilii* modo . . .'. His description of Caligula's use for a banquet of a plane-tree fitted out for the purpose (12. 10) is not condemnatory of the practice itself but pokes fun at the disproportionate size of the emperor for the occupation of what he called his *nidus*, in which 'ipse pars esset umbrae'.[53] The vine-shaded bedroom in the Younger Pliny's Tuscan villa (*Ep*. 5. 6. 39) is surely a simple imitation of the pleasures enjoyed by Mucianus; his uncle is clearly impressed by the vines to be seen climbing all over certain houses, a natural-looking display of spontaneity, but instigated and controlled by man all the same (14. 11).[54]

Next, we may consider another feature of the ornamental garden, *nemora tonsilia*, trees whose size and shape are artificially changed by man, giving them both an appearance and a function that were not intended by nature. To Pliny, this is not

[53] Verginius Rufus called his retirement villa his *nidulus* (Pliny, *Ep*. 6. 10. 1).

[54] Cicero (*QF* 3.1) describes a lavish display of ivy on the walls and between columns—perhaps more obviously controlled, as the Greek statues look as if they were doing ornamental gardening. The utilitarian vine could be approved as a provider of shade, as in Martial 12. 31 (contrast the barren plane of Martial 3. 58. 3). Cf. n. 56.

development of propensities already inherent in nature, but an arbitrary perversion—'*coactae* brevitatis' (12.13, cf. 16.140 '*coercita* gracilitate'). It is significant that he calls it *abortus*, another example of the violent interruption of a natural cycle which he criticizes elsewhere (Ch. VI.1.c). In fact, it is noticeable that he can refer to ornamental gardening (*topiaria opera*, 35.116), generally without a condemnatory tone (15.130; 16.70). Only in two instances, the pine (16.40) and the cypress (16.139), is there a hint of doubt. Significantly, both are called funereal trees. The clipping of these may well have offended religious scruples, especially as grafting fruit is forbidden in the case of some varieties by such taboos (*neque . . . fas . . . est*, 15.57).

Why is there a rather uneven tone of condemnation in Pliny, even when perversion and not arrangement is evident in man's interference with *Natura*? To some extent, it is a question of degree. Moralists are certainly concerned with the artificial redirection of specific natural objects from their usual form (*nemora tonsilia* in Pliny) or cycle of growth (flowers out of season in Sen. *Ep*. 122.8 or the evergreen plane in Pliny *HN* 12.11–12 and Sen. *Ep*. 122.8). A passage in Seneca the Elder (*Contr*. 2.1.13), however, is concerned with the extent of the alterations to which a man's arrogance and deluded self-importance can lead him. Since Pliny's comments on gardens are in the context of an examination of the individual plants, small specific organisms, the scope for man's arrogance is rather narrow. The huge landscape upheavals of the very rich are outside Pliny's subject. Otherwise we should expect him to have much to say on the gardens of the Domus Aurea, affording as it would an opportunity for another attack on his *bête noire* Nero. However, man's arrogance and self-delusion in his attempt to master *Natura* on an even larger scale, the cosmic, is a subject of some importance to Pliny, as was seen in Ch. I.4. Bearing all this in mind, it is perhaps less surprising to find that Pliny is comparatively uninterested in strong condemnations of small-scale ornamental interferences. Indeed, it would seem from archaeological evidence that such ornament might be used on a very modest and unpretentious scale.[55]

[55] See W. J. Jashemski on house 1. XII. ii at Pompeii (in E. MacDougall and W. J. Jashemski (eds.), *Ancient Roman Gardens*, Dumbarton Oaks. Colloquium on the History of Landscape, Washington, DC, 1981).

Another, and in Pliny's case, stronger reason is the principle of true value mentioned in the last section. Luxury as luxury is not as objectionable as that which takes on unnatural importance by impinging on the basic needs of life. Applying this principle to gardens, it is permissible to have a luxury ornament so long as it is recognized as such and no more. One should not, on the other hand, pay vast sums for a luxury version of a normally cheap and basic food. Thus Pliny's one real outburst against a garden luxury is connected not with an ornamental garden but with the products of the traditional utilitarian Roman *hortus*. Once, such a garden was the poor man's farm (19. 52, cf. Martial 3. 47) and the food produced was sufficient both 'voluptati satietatique'. Yet even here, Pliny complains, we meet with the same *indignatio* as everywhere else. In spite of all the other edible luxuries that exist, even the humblest vegetables must now be graded and certain types of specially fattened kale are grown for the rich (19. 54, cf. 19. 139–40 on luxury cabbage requiring twice the usual outlay and trouble: 'impendio taedioque' . . .). *Indignatio* carries a reproach of injustice and ill-treatment of the earth (cf. 33. 1 ff., Ch. I. 3) by the perversion of her basic gifts intended for all men, to fit man's artificial scale of wealth.

We can now reach some conclusions about the Roman concept of artificiality in general, and in particular as related to Pliny's ideas. We have seen that the control and ordering of *Natura* by man is not necessarily artificial in a bad sense. Man and *Natura* can, as in grafting, form a partnership to further her purpose. Even when large-scale interference against *Natura*'s intentions is criticized, emphasis is on man's delusive self-importance rather than his perversity *per se*. Such is the effect of the Roman tendency to view Nature primarily from the 'human' rather than the 'natural' standpoint. Furthermore, Pliny suggests that there is a friendly rivalry rather than a hostile competition between Art and Nature (21. 2–4). Not even (*ne . . . quidem*, 21. 2) the painter's art can copy the colour and variety of *Natura*'s flowers. The competition, 'certamen artis ac naturae' between the painter Pausias and the garland-maker Glycera is friendly (21. 4). Above all, the ambiguity is encapsulated in the basic idea of *Natura artifex*. Both Pliny and Columella (*De Re Rust*. 10. 96) describe Nature as painting the colours of the flowers. Art and Nature are overlapping not opposing spheres; there is art in nature. Distinctions become blurred. As mentioned in the introduction to this chapter, Pliny's books on art arise naturally out of an interest in the raw material provided by nature.

In fact, the idea of art in nature leads Pliny to make some very striking similes. He uses comparisons with ornamental gardening to describe wholly natural features and landscapes. In *HN* 4. 29, Mount Nymphaeus in Pthiotis is described as 'quondam topiario naturae opere spectabilis'. In *HN* 12. 22, the natural propagation of the Indian fig-tree is compared to ornamental gardening. The branches lean down to earth and take root, 'and they produce new offspring for themselves in a circle around the parent tree *quodam opere topiario*'. Moreover, Pliny goes on to describe the attractive appearance of this natural arrangement from different viewpoints: 'It is a pleasing arrangement with its vaulted dome when viewed from below or from afar.' The features of formal gardens, too, were planned with both close-up and distant viewing in mind. Pliny the Younger's *hippodromos* at his Tuscan villa allowed for an overall view (*Ep*. 5. 6. 32) of the whole course as one entered as well as paying attention to the variety in the immediate surroundings as one traversed the walks.[56]

This ambiguity of *Natura artifex* modifies Pliny's unfavourable opinion of even the most artificial and ornamental of gardens. If we turn from these to the traditional Roman *hortus*, we find that it too can influence the blurring of the nature/artifice distinction. Such a garden is Pliny's main interest. It is closer to his moral ideals, being basically utilitarian—a produce garden—and also playing its part in the tradition of early Roman *mores* (19. 50 ff.). However, the informal garden was not a thing of the past; gardens planted in this old-fashioned manner with trees and shrubs were not unusual in the last years of Pompeii, for

[56] A striking fascination with the idea of the competition between art and nature in the garden is to be found in the 4th-c. writer Gregory of Nyssa. In *Ep*. 20, he describes a visit to a garden at Vanota, near the Cappadocian–Galatian border. If one looks for natural beauty in this garden, it needs none of the adornments of art. If one considers what has been done for it by artificial aid, so much has been done, and that so well, as might overcome natural disadvantages (§ 5). In grafting (§ 11), nature, 'governed by art', turns one tree into an almond, another into a walnut, like a painter who subtly beguiles nature by inventing centaurs and other mythical beasts. In the garden, such competition enhances rather than devalues nature, cf. § 12, in the arrangement of trees, nature is 'willingly' restrained by man. In common with other writers on gardens, Gregory appreciates shade: he admires a path under climbing vines (*Ep*. 20. 13). Cf. n. 54.

example.[57] One imagines that the majority of gardens must always be informal ones, catering for both necessity and enjoyment.[58] It is therefore very likely that the informality of the ordinary *hortus*[59] blurred the distinction not only between natural and artificial order, but even between what is cultivated by man and what is not.

Dealing first with the question of order in such a garden; despite the laying out of well-manured beds (19. 60), the overall look of the garden must often have been one of spontaneous growth. Thus (19. 167–8) Pliny lists plants which are grown in company with others, such as poppy with cabbage. Cucumbers and gourds have a number of uses, some fruits being left to grow freely for eating, while others are forced into strange shapes by means of wicker sheaths (19. 70 ff.). In addition, they, like the vines mentioned earlier, may climb right up the house walls to the roof. This suggests that they were not confined to a particular spot, but were left to run wild, performing a further, ornamental role of vigour and spontaneity. Again, the planting of certain flowers would reflect practical as well as ornamental uses, as do the colourful plants with medicinal value in *HN* 22. 16, random clumps for practical use featuring as much as more ornamental displays. This could be the case when growing flowers for bees

[57] 'Old-fashioned' gardens: see Jashemski (n. 55) and her *Gardens of Pompeii*, i (New York 1979) for a shop's large informal produce-garden with trees, vines, and vegetables (1. xx. 5).

[58] See N. Purcell, 'Town in Country and Country in Town', in E. B. MacDougall and W. J. Jashemski (eds.), *Ancient Roman Villa Gardens* (Dumbarton Oaks, Washington, DC, 1987), 187–203.

[59] Certain types of garden are beyond the scope of this discussion. Large-scale commercial gardens were more organized and might specialise (Varro, *RR* i. 16. 3; i. 23. 4). There had been recent commercial developments in market-gardening (Colum. *De Re Rust.* 10, pref. 1), and Pliny's lack of interest might be due to moral disapproval of *quaestus* (*HN* 19. 47). But see sect. 4 above, and Ch. V. 2. B on his generally more flexible attitude to trade; note that 19. 47 refers to a species of madder, a type of dye (see n. 52). It is more significant that his only detailed remark on any aspect of commercial gardening is his objection to its production of 'luxury' vegetables, as discussed earlier in this section. Pleasure gardens are mentioned only once, in disparaging tone, as rivalling in size the entire amount of land (2 *iugera*) allotted to the individual for his livelihood in the early days of Rome. *Hortuli*, on the other hand, if small plots of land held individually (P. Grimal, *Les Jardins romains à la fin de la République et aux deux premiers siècles de l'Empire*[2] (Paris, 1969,) 57–60) would display the individuality, variety, and informality of the 'ordinary' *hortus*.

(21. 70 ff.) as not all such varieties were particularly decorative.[60] This impression of ordered spontaneity is well illustrated by Virgil's description of an informal garden in *G.* 4. 125 ff. The vegetable-beds are edged with flowers but the overall picture is of a show of colour and fertility.

Secondly, and even more significantly, there is a lack of clear distinction between the cultivated and the wild plant. One factor must be the medicinal properties of many wild plants which would ensure their place in the majority of gardens as well as in the wild. Some of these would, no doubt, be included only in specialist medical gardens.[61] Others, however, do not appear to be limited to specialists. In *HN* 19. 172 Pliny says that, in Greece, wild thyme and mint are picked from the mountains (for bees perhaps, see Virgil, *G.* 4. 112).

Lack of clear division is suggested when, after announcing that he is turning from gardens to discuss plants that grow 'sponte' (21. 86), he talks of asphodel being planted—*satum*—before the gate of a villa to ward off sorcery (21. 108). Indeed, plants that grow *sponte* are interesting for their ambiguity. Perhaps the most interesting passage with regard to this opposition between wild and cultivated is *HN* 20. 204 on the different types of poppy. Cultivated (*sativa*) and wild (*silvestria*) poppies are here distinguished from a third kind which grows 'sponte' on cultivated ground. Pliny goes on to describe its medicinal value. Many plants which we should call weeds did in fact have such uses, a fact which may explain why the Latin term for weed is *herba inutilis*. Yet, even thus defined, weeds were not wholly banished from the garden. Of course, garden-beds were weeded to get rid of plants harmful to the sown vegetables (Colum. *De Re Rust.* 10. 149, 11. 3. 13). Columella, however, sees some of these self-sown plants as sufficiently integral to the garden flora to be included with them in his calendar of growth (*De Re Rust.* 10. 373). It may be relevant to this lack of a clear dividing-line that an unqualified *herba* is often used when talking about weeds, as in

[60] Virgil, G. 4. 112: 'thymus pinosque . . . de montibus altis . . . '. See also below, on 19.172.

[61] *HN* 19. 167, a wild plant 'also grown in gardens, but only medical ones'. Medicinal plants were probably a prime feature of Antonius Castor's *hortulus* (25. 9), cf. references to plant remedies of his (20. 244, 261; 26. 51), and see Ch. VI. 2. D.

Cato, *De Agr*. 34–5, 48. 2 (cf. *herba topiaria* of cultivated plants, *HN* 21. 68, 22. 76).

There is similar ambiguity in the case of asparagus. Columella mentions wild asparagus amongst the wild plants which grow in gardens (*De Re Rust*. 10. 375). In *HN* 19. 54, Pliny complains that a luxury asparagus is grown for wealthy men who are not prepared to be content with the wild variety *Natura* had provided for everyone. In *HN* 19. 145–6, he describes another type, *incultius* than the normal garden-grown variety, but *mitius* than the wild one. According to the emperor Tiberius, it was 'simillimam asparago' yet it grew in Upper Germany like a weed, *herba*. Indeed, adds Pliny, that which grows *sponte* in the island of Nesita off Campania is considered to be by far the best type of asparagus.[62]

The traditional Roman *hortus* was not, therefore, clearly divorced from 'wild' nature.[63] The art of man and the art of Nature need not be at odds, especially in the utilitarian rather than the ornamental setting which such a garden provides. It remains to examine the implications of describing *Natura* as *ludens et lasciviens* in such a setting. It is, of course, partly an expression of the sheer beauty of Nature without thought for ends and purposes. A phrase used in an anecdote about sums paid for beautiful trees (17. 5) echoes the language in *HN* 21. 1: lotus-trees are 'patula ramorum opacitate *lascivae* . . .'. Yet Pliny's ever-present practicality even with regard to flowers a little further on in *HN* 21—their colours signify their medicinal value—bids us look further. If certain words used in descriptions of informal produce-gardens are examined and related to the

[62] The self-sowing of useful wild plants would be greatly approved by those who considered that luxury types were cultivated unnaturally because unnecessarily. In contrast, the sprouting of wild varieties *sponte* in the garden as well as outside it would reinforce the idea that the garden was a harmonious partnership between *Natura* and man. Seneca *Vit. Beat.* 17. 2 sums it up: 'Quare cultius rus tibi est quam naturalis usus desiderat?' Syme ('Diet on Capri', *Athenaeum*, 77 (1989), 261–72: *Roman Papers*, vi, ed. A. R. Birley (Oxford, 1991), 409–20), identifies the *herba* of *HN* 19. 145–6 as the hop, following J. André's Budé edition of *HN* 19, many plants with edible stalks and leaves being termed *asparagus* by the ancients.

[63] The blurring of the nature–garden distinction is also to be seen in an interesting description from the east of a sort of semi-garden. Clement of Alexandria says that his *Stromateis* ('Miscellanies') are not laid out like a formal

phrase in *HN* 21. 1, they will be seen to produce a coherent picture of the ideal relationship between man and *Natura*.

The use of *fertilitas* in *HN* 21. 1 produces the key. Virgil (*G*. 4. 118) used the epithet *pinguis* of the *hortus*; Pliny the Younger talks of a garden *pinguis et rusticus* (*Ep*. 2. 17. 16). *Rusticus* reinforces the arguments of the last section that divisions between gardens and all but the very wildest countryside were vague. *Pinguis* gives a basic reason for seeing the garden in particular as a microcosm of the fertility of *Natura*. When Nature was looked upon from the point of view of man, her greatest gift was the fertility of the earth as a whole. It is noticeable that the agricultural writers emphasize using very good soil for the *hortus*: Cato (*De Agr*. 1. 7) puts the garden second in his list of nine types of farming, and Varro (*RR* 1. 7. 9), commenting on this, specifically relates these categories to grades of soil, disputing only the choice of vines rather than pasture-land for the top grade. With regard to Varro's preference, it is interesting that Columella should recommend for gardens (*De Re Rust*. 10. 6 ff.) *pinguis ager* where grass flourishes. We may compare Virgil, 'tortusque per herbam / cresceret . . . cucumis' (*G*. 4. 121–2). One thing which made the particular garden which Virgil goes on to describe so remarkable was the fact that, in spite of the land's not being fertile enough for any type of agriculture, its owner still managed to raise a small but luxuriant *hortus* (*G*. 4. 128–9).

Fertilitas in *HN* 21. 1 is qualified by *varia*. *Varietas* is one of Nature's most important attributes in Pliny and is linked with the *ludens* image in several passages.[64] It occurs in many of the writings on gardens. Cato talks of 'hortum omne genus, coronamenta omne genus'. . . (*De Agr*. 8. 2), combining flowers and vegetables in the garden of the suburban villa. Columella's first epithet for the garden is *numerosus* (*De Re Rust*. 10. 6). Like Pliny, he sees flowers as a striking manifestation of *Natura's varietas*: 'varios, terrestria sidera, flores' (*De Re Rust*. 10. 96).

garden, where everything is planted in orderly fashion to please the eye. Rather, they are like a shady hillside, wooded with various trees in natural irregularity, among which the gardener conceals his fruit trees to hide them from thieves. He also uses the hillside as a nursery from which to take plants and cuttings for the formal garden (*Strom*. 7. 18. 111. 1).

[64] See sect. 4. For *varietas* in connection with *natura ludens*, see 9. 102 on sea-shells and 11. 123, 'nec alibi maior naturae lascivia; lusit animalium armis', referring to the variety of horns with which she has equipped the beasts.

Variety on a larger and more generalized scale—different types of garden, and, further afield, a varied view of the countryside beyond—was an important feature of larger estates.[65] The traditional *hortus* is, however, small enough to be an easily apprehended microcosm of the most basic life-giving properties of *Natura*, expressed by the phrase *fertilitas varia*.

It follows that *lasciviens* and *ludens* also denote fertility. The use of *gaudium*, too, in such a context, may be paralleled by Virgil's 'potis gauderent intiba rivis' (*G.* 4. 120) of endive flourishing in a garden by a stream and by Columella's 'alma parens . . . gaudet' (*De Re Rust.* 10. 257). Pliny's description of the lotus-trees (*HN* 17. 5) mentioned earlier suggests that their beauty is due to the exuberance of their natural growth. *Laetus*, too, is often used of thriving crops, as for instance in the beginning of Virgil's first *Georgic*: 'Quid faciat laetas segetes . . . canere incipiam'. Cato uses it of rich soil (*De Agr.* 61. 2). In fact, all these words used of plant growth and suggesting joy have the same connection with luxuriant fertility as does the literal meaning of the English 'exuberance'. Pliny describes the Campanian shore as 'felix . . . ac beata amoenitas . . . gaudentis opus . . . naturae' (*HN* 3. 40), and the whole surface of the earth as rejoicing: 'superficies [*soli*] caelo quocumque gaudens' (19. 56). *Natura*'s fertility is the source of man's existence. Thus the pleasure she shows here in her own gifts is also a sign of her benevolence towards him. In this sense, of *Natura provida*, *Natura ludens* is a valid and unambiguous contrast to *Natura saeviens*.

Finally, the 'interference' of man with *Natura* is not simply the most harmonious of partnerships; her productivity is actually increased through man's activity. The old man in the fourth *Georgic* got more out of his patch of land than was possible under other types of agriculture. The *hortus* is a small area, cultivated intensively: *Natura* in the setting of a garden is, moreover, protected against her own harsher elements. These may be the natural laws of survival and competition; the garden is carefully enclosed against the intrusions of hostile men and beasts (Colum. *De Re Rust.* 10. 28), allowing even the weakest seedling to come to fruition. Even more significant is the fact that the ambiguous products of *Natura*, which raise doubts as to the

[65] See Pliny, *Ep.* 5. 6. 13 and Grimal (n. 59), 200.

extent of her benevolence towards man, are seen as alien to the garden setting. Columella (*De Re Rust.* 10. 17–18) says that soil which harbours poisons is unsuitable for a garden. The right soil 'refuses to harbour the hellebores with their harmful juice / nor does it allow yews to grow, nor does it exude strong poisons'. Not all poisons render the ground unsuitable (ibid. 19), but the contrast of this soil with the right type (11–17), so fertile in fruit, a basic food (12. 1; see intro.) makes the point sufficiently clear. We are left in no doubt as to the appropriateness of the garden as the epitome of *Natura*'s exuberant fertility in products beneficial to mankind.

III

Man and the Gods

1. *Religious Attitudes in Pliny*

The only form of deity acceptable to Pliny is *divina Natura*, the spirit of the universe, an all-embracing unity. Traditional ideas of a multiplicity of gods and goddesses as in the Olympic pantheon are merely examples of *imbecillitas humana* (2. 14). The diversity of religious ideas only obscures the search for God (2. 22). Some men slavishly worship *monstra* with *externa sacra*, while others believe in nothing at all (2. 21). Some invent a deity between these two concepts, *Fortuna*; others again adopt the fatalism of astrology. Pliny concludes that 'naturae potentia . . . est quod deum vocemus' (2. 27).

The severely sceptical line taken in these chapters does not prevent Pliny from scattering the *HN* with references to *dei* and *numina*.[1] These need not, however, be seen as inconsistencies; Pliny is simply using the traditional modes of expression for what are casual references to the outstanding, the extraordinary, and the 'superhuman'. The intention to use the term *deus* in describing pinnacles of human achievement had, of course, been stated by Pliny in 2.18 ('deus est mortali iuvare mortalem . . .'). Thus in 2. 192, for instance, the predictors of earthquakes may be regarded as equivalent to gods. The idea that 'only a god' or 'not even a god' could achieve a particular feat occurs frequently. A number of these have already been discussed in different contexts. We have seen (Ch. I. 4) that Hipparchus' attempt to enumerate the stars was a rash deed 'etiam deo improbam' and that drug-mixers falsely claim a minute accuracy which can only have been dictated by some god. Similarly, mysteries of nature are sometimes said simply to be the action of a god, when Pliny believes them to be beyond human rationale (Ch. II. 2). Thus 'a

[1] e.g. 2. 192; 25. 16–17; 27. 2–3, 6, 8; 37. 59–60.

god' discovers the rabies cure revealed to a woman in a dream (25. 16–17). In one instance, the ability of goat's blood to break diamond, the discovery is so bizarre (37. 59–60, cf. Ch. VI. 2. c) that Pliny is reluctant to attribute it even to *ratio naturae*. *Deus* and *dei* can also be straightforward synonyms for *Natura*: plants, the gifts 'of the gods' are given to man by the generosity of *Natura* (27. 1–2).

What we have in such passages is a convenient and easily comprehensible shorthand method of describing the superhuman, rather than a serious lapse in Pliny's avowal of scepticism. It is only to be expected that an educated but general writer like Pliny might juxtapose his formal analysis of *deus* with informal everyday usage of the same term.

Unlike his casual references to *deus*, Pliny's statement in *HN* 2. 14–27 is very self-conscious. He takes pains to justify his digression. 'It will not have been irrelevant to have diverged to these topics which have already been widely disseminated because of the unceasing inquiry into the nature of God' (2. 27). His personal conclusion, that Nature is God, is, of course, anything but irrelevant to his portrait of the universe as a whole. It remains to review the wider question posed by his rejection of the numerous beliefs which this unceasing search for God has inspired. In order to explain his extreme scepticism, it is necessary to decide what he was rejecting and why. The later part of this chapter will examine individual ideas criticized by Pliny in the *HN*. The first task must be to examine the nature of the religious debate suggested by Pliny's justification in *HN* 2. 27, and then to set out the main lines of inquiry and criticism followed by Pliny in his attack on religion.

Upsurges in religious interest have traditionally been associated with periods of political turmoil in the ancient world. The spread of religious ideas of Eastern origin in the Hellenistic era was encouraged by developments in the political situation. These included not just the political expansion itself but also the effect that insecurity or lack of personal control over material well-being had on individuals.[2] Mystery religions could provide men with divine saviours in misfortune: *Tyche* became personal as well as

[2] A. J. Festugière, *Personal Religion among the Greeks* (Berkeley, Calif., 1944), 39 ff.

public.[3] The increasing control of the Roman empire by a single ruler would tend to promote feelings of personal powerlessness. Astrology and magic offered an alternative means of controlling not only the individual's own destiny but also that of others. More recent theories have undermined any idea that the resulting religious atmosphere at Rome was one of opposition between rigidly compartmentalized 'old' and 'new' ideas and rituals.[4] The Roman religious attitude was amenable to the assimilation of novel elements; their growth was not necessarily a sign of disillusionment with and opposition to a state ritual devoid of spiritual meaning. The traditional rituals continued to serve their purpose as a guarantee of social and political stability. Political instability in the Late Republic encouraged attempts to preserve and explain that ritual as well as inquiries into new ideas.[5] This is worth bearing in mind as it might seem easy to explain away Pliny's forthright condemnation of magic, astrology, *Fortuna*, and *externa sacra* as being reprehensible foreign accretions to traditional Roman beliefs.

Pliny's emphasis on the 'unceasing inquiry into the nature of God' also reflects a specific intellectual interest in matters of religion. The influence of Greek philosophy had done much to encourage both the exploration of newer ideas and the interest in older Roman practices in the Late Republic. Pliny's own emphatic scepticism and the growing intellectual interest in various religious ideas which he records are themselves products of this development. Indeed, his criticisms are to no small degree a defence of his own intellectual position against rival ideas.

The lines of inquiry and criticism followed by Pliny are determined in the main by those beliefs described in the first two chapters. He believes in a deity, *Natura*, but she is kept vague by her pantheistic nature. His view of life springs from a mainly

[3] For *Tyche*'s ancestry and her relationship to the Roman *Fortuna*, see the article in Daremberg–Saglio ii. 2. 1264–77 (Paris, 1896) on Fortuna–Tyche by J. A. Hild; J. Ferguson, *Religions of the Roman Empire* (London, 1970), Ch. 5.

[4] J. A. North, 'Conservatism and Change in Roman Religion', *PBSR* 54 (1976), 1–12.

[5] E. Rawson, *Intellectual Life in the Late Roman Republic* (London, 1985), 298–9. The wide interests of Pliny's scholarly predecessor, Varro, included types of divination (Rawson 314) and, according to Pliny himself, he was buried with Pythagorean rites (35. 160). Cf. also, Nigidius Figulus, one of Pliny's most frequently quoted sources (indexes bks. 7, 8, 9, 10, 11, 16, 29, 30), called 'Pythagoricus et magus' by Jerome (*Chron*. 45 BC, Rawson 292).

Stoic outlook. Within the framework of *Natura*, man, the supreme creation, must live independently and positively. He should exploit the gifts of Nature, both those around him and especially that within him, *ratio*. The attack on certain religious ideas is based on their inherent opposition to facets of this outlook. They may degrade the supremacy of man in his use of *ratio* and his positive independence in facing life. They may also degrade the power and sanctity of *divina Natura*.

Pliny arranges his material in *HN* 2.1–27 so as to contrast the *potentia* of the all-embracing unity of *Natura* with the divisions of opinion created by *imbecillitas humana*. This is an effective way of stressing the superiority of his own view. However, it should not obscure the fact that he opposes beliefs old and new as a body. The objects of his criticism are common to all the beliefs he mentions individually, though there may be differences of emphasis and detail from one to another.

We may start with the supremacy of man. Pliny's view of the life of man in nature is above all positive (Ch. II, intro). Despite the manifest power of *Natura*, she is still, in many respects, the servant of man (2.157 '[*terra*] servit homini'). Pliny, as has been seen (Ch. II.3), portrays fear as a denial of man's rationality. It follows that religious practices which involve a servile 'fear-cult' of gods are to be condemned. In *HN* 2.21, Pliny describes scenes which are very close to those criticized by Plutarch in his *De Superstitione*. Pliny describes men who are 'slaves to foreign rituals', who 'carry gods on their fingers' (engraved on rings), and 'curse the monsters they worship'. They subject themselves to 'terrible tyrannies' and cannot do anything except after sacrifice. His opposition to fearful placation leads him to be equally critical of certain ancient Roman practices such as the deification of virtues, vices, and diseases, which has come about 'because of our fearful anxiety to appease them' (2.15–16).[6] In *HN* 2.18 ('deus est mortali iuvare mortalem') he had suggested that human co-operation was the most important 'divine' idea. Indeed,

[6] M. P. O. Morford, *The Poet Lucan* (Oxford, 1967), Ch. 5, sees a contrast in Lucan's *Bellum Civile* between the man who gives in to weakness and the Stoic who stands alone. The picture of Cato, who refuses to consult the oracle of Ammon (9.544–86), contrasts with the divination episodes involving the fearful people of Rome (1.673), the coward Appius (5.68–70), and Sextus Pompeius, who is driven by fear to discover the future.

such practices are for him more of an insult to man than, as in Plutarch, to the deity: 'It is hardly appropriate to decide which is more profitable for the human race, when some pay the gods no respect at all and others pay regard of a shameful nature' (2.20–1).[7]

Plutarch's definition of superstition is in a specialized sense of *deisidaimonia* as 'fear stemming from ignorance'. It may still be related to Pliny's general fear that human capacity for rational thought is being ignored by such beliefs. The tendency had been caricatured in an earlier age by Theophrastus: if the magician says that a bag gnawed by a mouse is to be patched, the superstitious man will ignore him and concentrate upon apotropaic rites instead (*Char.* 16.6). On a more serious note, while Pliny appreciates the problems of physical and emotional weakness (Ch. II.3), he is also confident that physical disadvantages can be overcome by an individual to a quite remarkable degree. M. Sergius Silus, the great-grandfather of the conspirator L. Sergius Catilina, persistently overcame the crippling effects of wounds to fight in numerous battles (7.104–6, below, 2.c). Even more significantly, the psychological afflictions like fear may be vanquished by *ratio*, especially if one man uses his talents to help his fellow men as in the example of the astronomer Hipparchus (2.53–5, Chs. I.4, II.3). It is significant, therefore, that Pliny's criticism of the move towards astrological belief is couched in terms which suggest the very opposite of rational co-operation. 'This belief begins to take root and the learned and the unlearned mobs alike go marching on towards it at the double' (2.23). Here, even the educated,[8] better equipped to be rational and

[7] Cf. Plutarch, *De Superstitione* 170 E. Pliny does see the gibberish of magical formulae as an insult to the being thus invoked, though this does not mean that he necessarily believes that an invocable deity exists. One expects 'something huge, something fit to move a god or, rather, to give orders to divinity' (*HN* 28.20).

[8] J. Thérasse, 'Croyances et crédulités des Romains d'après Pline l'Ancien et les écrivains latins', in J. B. Caron, M. Fortin, and G. Maloney (eds.), *Mélanges Lebel* (Quebec, 1980), 283–319, claims to show the prevalence of superstitions among the more educated classes, including men in high office and men of letters. As far as Pliny's 'credulity' is concerned, however, too much stress is laid on his inclusion of many beliefs, although some allowance is made on the evidence of *HN* 11.6: 'For when I have observed nature, she has always induced me to consider no statement about her incredible.' Not enough weight is given to the non-committal way in which many of the beliefs are reported by the writer. Moreover, the significance of such passages as 2.23, where Pliny implicitly dissociates himself from the credulity of his contemporaries, both unlearned and

help others, join in a mindless, panic-stricken rush. Similarly, the phrase used of those who reject *Fortuna* in favour of astrology—'pars alia et hanc *pellit*'—suggests a lack of any rational decision (ibid.). Finally, to submit to fatalistic beliefs, whether it is the certainty of astrological predictions or the 'certain uncertainty' of *Fortuna*, is to reject the active independent stance in life appropriate to the possessor of *ratio*.[9]

'The learned and unlearned mobs alike' encapsulates not only the negation of man's rationality but also that process of blurring the distinctions between more and less respectable ideas. Such learned interest in some of the ideas Pliny criticizes is another facet of the abuse of *ratio* which he opposes. Intellectual activity is of prime importance in Pliny's view of *vita*. In the days before learning was stifled by avarice, 'the rewards and achievements of life abounded' (14.4–5). But now, magic in particular had aroused educated interest. Clothed in an intellectual aura, it could claim the same respectable standing as could Pliny's beliefs in *Natura*. The latter were based in the philosophical ideas which had developed in the late Republic. It may well be in the same period that intellectual interest in magic's association with Eastern learning had started.[10] Magic had, of course, long been embedded in popular religion, but its elevation to an *ars*[11] gave it a new and, to Pliny, spurious respectability.

So far, we have seen that Pliny's opposition to certain religious ideas is connected with their abuse of man's dignity and intellectual independence. These qualities are undermined by servile cult-practices, by passive dependence on the fatalism of astrology, and by the opposite extreme, submission to the uncertainty of *Fortuna*. The claims of magic to intellectual dignity are dangerous because they are false. All are at odds with Pliny's idea of *vita*: positive living and the sovereignty of mind of the

learned, is missed. So too are his interesting remarks about the ambivalent attitude shown by some of the more educated (28.10 ff.) and by the powerful (2.94; 8.117; 14.88). See below, this section.

[9] Compare Plutarch's outlook in *De Fortuna* 98 E–F: to attribute everything to *Tyche* is an insult to man's superior intelligence.

[10] J. A. North, 'Novelty and Choice in Roman Religion', *JRS* 70 (1980), 186–91; cf. A.-M. Tupet, *La Magie dans la poésie latine* (Paris, 1976), 419–20, and R. Garosi, 'Indagine sulla formazione del concetto di magia nella cultura romana', P. Xella (ed.) in *Magia: Studi di storia delle religioni in memoria di Raffaela Garosi* (Rome, 1976), 13–93, at pp. 54 ff.

[11] Rawson (n. 5), 309.

individual. These same principles hold true in the case of Pliny's attitude to traditional Roman ritual. In this area, perhaps, it is even clearer that the true dignity of man the rational being matters more to him than the literal existence or otherwise of the state gods.

It has been noted that the formulaic prayers of the Roman state religion are suggestive of an agreement made between rational beings.[12] It was thus an aspect of religion fitting to the dignity of man as well as being of importance in ensuring the security of the state. These would seem to be the ideas in Pliny's mind when he says of formulaic prayers that 'if one reads (them), one is forced to admit that there is power in prayers, the events of 830 years showing this for them' (28. 12). The solemn and dignified formulae are impressive; they have served Rome well over the ages. Whether they can actually move gods or not remains an unconfirmed hypothesis. 'If this view is once admitted [that Vestals can root to the spot runaway slaves by incantations], and that the gods hear certain prayers or are moved by any form of words, the whole question must be answered in the affirmative' (28. 13).[13] What matters is the attitude and the intention of the person who performs such ritual. In a discussion on the significance of omens, he concludes that 'these instances suffice to show that the power of omens is really in our control and their influence depends on the way we receive each one' (28. 17). He goes on to mention the augurs' precept that evil omens do not affect those who declare beforehand that they take no notice of them. In religious observances man should not regard himself as a passive slave. He should be tyrannized no more by tradition, formulae, and ritual than by foreign *monstra*. Pliny approves of the stand taken by M. Sergius (above), who, when his fellow praetors wished to debar him from the sacrifices as being infirm, objected with a speech listing the exploits in which he had, on behalf of his country, incurred his injuries. It would be interesting, in the context of Pliny's argument and for the wider question

[12] North (n. 4).

[13] Th. Köves-Zulauf discusses this and other remarks on the power of words in his exhaustive study of the religio-magical significance of words as seen in the *HN*'s material. He concludes that Pliny's personal attitude is at least positive, whether or not he actually 'believes'. *Reden und Schweigen; Römische Religion bei Plinius Maior* (Munich, 1972), esp. 29–34.

of flexibility in ritual to know whether Sergius was able to overcome his colleagues' opposition or not.[14]

Certain religious attitudes are, then, opposed to Pliny's ideal of man. They erode his dignity and distort his rationality, either by fear or by false rationality as in the 'intellectual' approach to magic. It should, however, be noted that he is aware of the complexity of the psychological influences of popular beliefs. He notes a difference of attitude between the intellectual and the non-intellectual, and between the individual and the crowd. On trivial superstitions of everyday life, he says that, as individuals, the wisest men (*sapientissimi*) reject them, but people generally believe in them, if unconsciously (28. 10 ff.). The more literal and crude approach of the people *en masse* can be channelled to good effect by individual leaders. In *HN* 14. 88, he describes how King Numa decreed that no wine-offerings could be made from unpruned vines 'in order to compel the people not to neglect pruning'. In *HN* 2. 94, he distinguishes between Augustus' 'public' interpretation of the comet which appeared after the death of Caesar, and his private feelings. In *HN* 8. 117, he says that Sertorius 'persuaded' the tribes of Gaul to believe that his unusual white hind was prophetic. Because of the inseparability of Roman religion and politics, political leaders were the chief arbitrators in religious matters.[15] In Pliny's view, popular superstition is thus under the control of those with the appropriate authority and intentions. All is properly ordered and rational, in contrast with the picture of seething instability which Pliny scathingly creates in *HN* 2. 22 ff. There, one section of the human race is under the volatile rule of fickle fortune, while another rushes headlong to embrace astrological predestination.

Finally, we may turn to those religious attitudes which are

[14] M. Sergius Silus was praetor in 197 BC (Livy 32. 27. 7; 32. 28. 2; 33. 21. 9; 33. 24. 4). He was probably cited as an example of *virtus* by Varro (R. Schilling's Budé edn. of *HN* 7 (Paris, 1977), 178) and the story was later used from Pliny by Solinus. There is, however, no evidence about his speech other than what Pliny tells us, cf. *ORF*[2] 97. On tradition in ritual, see North (n. 4). The following precept is quoted by the elder Seneca: 'Sacerdos non integri corporis quasi mali ominis res vitanda est' (*Contr.* 4. 2), but *vitanda* is not as strong as *vetanda*. Mere participation in the rites, such as Sergius desired, may have been less rigorously controlled than appointment to a priesthood.

[15] M. Beard and M. Crawford, *Rome in the Late Republic* (London, 1985), 29–31. The senate rather than the priesthood was the chief intermediary with the gods and controlled men's behaviour with regard to them.

opposed more specifically to Pliny's ideal of *Natura*. These may involve ideas which challenge her power or which may be construed as an assault on her sanctity. For Pliny, *Natura* is supremely powerful and is also the only entity he is prepared to call God. Interference by man with *Natura* may be an act of arrogance against a superior power, an act that is laced with sacrilege since *Natura* is divine.

Many of the claims traditionally attributed to magicians in particular involve the breaking of natural laws and are in clear opposition to Pliny's views. Ovid puts a comprehensive list of these into the mouth of Medea in *Met.* 7. 199 ff. She claims that she has made streams flow backwards, calmed and stirred up the sea, summoned and dispelled storms, uprooted rocks and oaks, moved forests, shaken mountains, and caused the earth to groan.[16] She prefaces this account with an invocation to various natural forces: Night, the stars, the earth 'who . . . instructs magicians in powerful herbs', breezes, winds, mountains, streams, and pools (*Met.* 7. 192 ff.). The description of a magician in Pseudo-Quintilian effectively sums up the opposition of the magical to the Plinian outlook: 'A man is summoned whose profession it is to go against nature' (*Decl. Mai.* 10. 15).[17]

In a more general sense, magic is an attempt to master the *vis* and *maiestas* of *divina Natura*. Pliny's use of the abstract phrase *vis naturae* makes the *maiestas* of his divinity seem all the more awesome and indefinable. Stoic *pneuma*, closely related to Pliny's 'diffusae . . . per omne naturae . . . numen' (*HN* 2. 208, Ch. I. 1), is similarly vague and immeasurable. However, the growth of the idea of deity as an abstract 'power' was also

[16] 'et mugire solum manesque exire sepulchris!' (206). *Mugire* is expressive. Perhaps these two operations may be connected with each other, and earth groans as the ghosts are summoned forth. We may contrast Pliny's indignation at the upheaval of earth even by non-magical means in the mining (Ch. I. 3) of precious metals. Men 'marvel that occasionally the earth gapes open or trembles as if it were not possible that this may be an expression of the indignation of our holy parent. . . . We penetrate her inner parts and seek for riches in the abodes of the spirits of the dead' (33. 1–2). Earth hides luxuries for man's own good (cf. Hor., *Carm.* 3. 3. 49; Ov. *Met.* 1. 140). Her concealment of the dead is an even more basic service to him (2. 154).

[17] Tupet (n. 10), 18. She notes that this literary source is unusually comprehensive. *Defixio* tablets name some of these forces but not all together. She sees in Ovid a particular interest in the magician's desire to interfere with natural laws (416 ff.). For Ps.-Quintilian and a discussion of the same idea in Tibullus, see Garosi (n. 10), 61.

encouraged by the less reputable creeds of astrology and magic. A. D. Nock[18] saw this development as a result not only of syncretism with regard to the deities themselves but also of the influence of magic on the uses of words denoting 'power', such as *virtus*, *dynamis*, and *arete*.

Pliny is not likely to believe that the magician can ever realize his more grandiose claims to manipulate *vis naturae* (below, sect. 2. A. ii) since he believes man's *ratio* is inherently inferior to nature's. It is also obvious that he considers the links between the power of *divina Natura* and man to be less simple than was suggested by many commonly held beliefs, including magic. He is sceptical with regard to all direct influence between the divine and the human individual. A man's fate may be part of the overall pattern of which omens are also a part, but the omens do not 'cause' his fate (2. 97, see below, sects. 2. A. i, B). The cruder ideas about omens and causation are also implicitly criticized by Tacitus when he points to Nero's survival long after some of the 'downfall' omens.[19] Pliny's lost history is cited as a likely source for the bulk of these 'Neronian' omens,[20] but his motivation is more likely to have been his hatred of Nero[21] than acceptance of some of the beliefs of the *vulgus*. This is not to deny that some more general form of cosmic sympathy is consistent with his ideas on Nature.[22] Yet, as we shall see in the case of divination,[23] he is reluctant to place importance on particularized, trivial instances. The point remains that for Pliny there is no simplistic direct line from God to man, no direct tie, for instance, between a particular man and a particular star as the *vulgus* naïvely believes (2. 28, see below, sect. 2. A. i). That intercourse between the individual man and superhuman forces which the magician hopes to achieve is implicitly denied.

[18] A. D. Nock, 'Studies in the Graeco-Roman Beliefs of the Empire', *Essays* (Oxford, 1972), i. 37.

[19] *Ann.* 14. 12, cf. interpretations of omens by the *vulgus*, 14. 22.

[20] W. Liebeschuetz, *Continuity and Change in Roman Religion* (Oxford, 1979), 165 n. 7.

[21] See Introduction.

[22] Sympathy: e.g. 20. 1; 29. 61; 37. 59, cf. Ch. VI. 2. C. For Pliny, Nero is an enemy of mankind; even his birth was unnatural (7. 46). The deeds and demise of such a monster could well be complemented by other aberrations of Nature. However, the usefulness of such omens to Flavian political propaganda should not be forgotten either.

[23] See Ch. IV. 3. 1.

It is worth looking in greater detail at Pliny's assessments of certain popular beliefs, both significant and insignificant, as they provide apt illustration of the issues just outlined. They undermine man's dignity and his relations with *Natura*. They oversimplify the links between the human and the divine. Finally, their faults may be veiled by a spurious intellectual respectability which makes it all the more necessary for Pliny to challenge them.

2. *Popular Beliefs Featured in the* HN

A. ASTROLOGY AND MAGIC

(i) *The learned appeal*

The greatest challenge which Pliny has to face in combating the claims of astrology and magic is their appeal to a learned as well as to an uneducated audience. Refutation on a more popular level was relatively simple. Greater subtlety could be claimed for Stoic ideas of cosmic sympathy, for instance, as opposed to the simplistic man–star links of popular astrology. As we have seen (sect. 1), Pliny can point out that the stars are attached to the *mundus*, not assigned to each of us in the way that the *vulgus* believes (2. 28). There is no such close alliance between us and the stars that they share our mortality. In the case of the common beliefs about Augustus' comet, he will admit only that there might have been some subtler cosmic sympathy (see above, sect. 1) at work, in that 'it had a health-giving influence over the world' (2. 94). In *HN* 2. 97, it is made clear that stellar phenomena and the fates of man should be linked only in so far as they are each part of a bigger, subtler chain of events.[24] This basic point is enough for Pliny. No attempt is made to tackle the more demanding problems associated with fatalism and free will. Recognition of an intellectual gap between the vulgar astrological beliefs and the more sophisticated philosophical views is sufficient.

Adoption of astrology by the learned, however, obscures this intellectual gap. It becomes sophisticated. Blended with

[24] See above, sect. 1, *fin*.

respected learning, it acquires *auctoritas* and becomes difficult to isolate and dismiss. Astrology's links with learning were already present in its Oriental associations.[25] It could be regarded as a reputable science, as is suggested by the possibility of interchanging the words *astrologia*, *astronomia*, and even *mathematicus* when describing the astrologer and his art.[26] Many ancient scientists admitted astrology to their studies. Pliny tries to make distinctions. Hipparchus is praised for relieving human fear of the celestial bodies by the scientific process of predicting eclipses and not for anything we should class as astrology. The blend in Manilius' *Astronomica* illustrates the difficulty of making such distinctions. He includes mathematics and astronomy with astrological ideas and adds a Stoic religious tone which is at times identical with Pliny's. Even allowing for the lack of clear dividing lines in ancient thought, the intrusion of what we should consider to be deduction from recognized natural regularities can lead to questionable analogies and comparisons in astrological writings. The second-century writer Ptolemy, in his defence of astrology in the *Tetrabiblos*, combines the doctrine of sympathy with the observation of weather-signs to prove that one should therefore be able to tell a man's temperament from the conditions at the time of his birth. Even if the idea of a macrocosm–microcosm link aids his analogy between the temperament of earth and that of man, it still does not bear close scrutiny. Weather changes from day to day, but man's temperament is being forecast as a reasonably stable element throughout his life. Columella specifically contrasts the prognostications of the farmers and animals of Ptolemy's comparison with the over-precise astronomical predictions of the Chaldaeans. The less pretentious observations of the rustic allow for variations and imprecision (*De Re Rust.* 11. 1. 31–2).

[25] For which see F. H. Cramer, *Astrology in Roman Law and Politics* (Mem. Amer. Phil. Soc. 37; Philadelphia, 1954), 5–9; Rawson (n. 5) 306.

[26] For the nature of the connection, see Ptolemy, *Tetr.* i, *proem.* 1, where our 'astrology' is called 'the prognostic result achieved through astronomy'. 'If astronomy is the study of the movements of the heavenly bodies, then astrology is the study of the effects of these movements' (K. Thomas, *Religion and the Decline of Magic* (London, 1971), 337). Astronomers such as Hipparchus have astrological works ascribed to them. e.g. *Cat. Codd. Astr.* 5. 4, p. 209 (though see O. Neugebauer, 'Notes on Hipparchus', in *Studies Presented to Hetty Goldman* (USA, 1956), 292 ff., esp. 293–4).

The *Tetrabiblos* also provides an illustration of how astrological writers might seek to attract learned readers by trying to eliminate the crudities of the popularized form of the beliefs. Ptolemy takes much trouble to elevate astrology to the status of a learned art. In the introductory passage of the *Tetrabiblos* (1.2.9), the comparison of medicine and astrology is so phrased as to enhance astrology's status by association with a skilled field of learning (cf. Ch. VI. 2. B). It also suggests that the two *artes* have in common aims of such a high and subtle nature that human skill is stretched to its limits and will not always attain to exact correctness.[27] The latter amounts to an association with much of the traditional Graeco-Roman philosophy and its tendency to reach a reasoned uncertainty rather than declare a certainty of the type sought by the adherents of fatalistic creeds.

It is also significant that Ptolemy, in trying to portray astrology as a learned and rationally evolved doctrine, encounters a fundamental problem. There is a yearning for mystery and wonder-working, a dislike of methods which make use of rational faculties on a matter-of-fact level. Even his learned addressees, it seems, hanker after miraculous revelation. In *Tetr*. 1.2.9, he takes these people to task. We should not complain that an astrologer makes use of already existing accidental qualities, such as nationality, country, and rearing. There may have been a difference in the level of sophistication[28] between the astrology of the upper, more educated classes who could afford the costly process of casting horoscopes, and the simpler practices of the lower strata of society; yet there is a psychological factor common to all inquiries. It is, not surprisingly, Tacitus who works this into a trenchant generalization. In *Hist*. 1.22 ff., Ptolemaeus works out Otho's chances from 'coniectura et rumore', but Otho takes his conclusions as 'peritia et monitu fatorum praedicta.' The attitude

[27] A funerary inscription for a dead child curses the lies of the *mathematicus* (*CIL* vi. 27140) but *Cat. Codd. Astr*. i. 106–7 shows astrologers openly admitting failure with the excuse that some factor was not taken into consideration (refs. in F. Cumont, *The Oriental Religions in Roman Paganism* (New York, 1956), 168 nn. 14, 15).

[28] See Cramer (n. 25) 47 ff. on itinerant astrologers of the type against whom Cato warned his farm-hands (*De Agr*. 5.4) and whose modest fees mocked their claims to sophistication (Cic. *De Div*. 1.132). For the astronomical and mathematical skills needed to cast a horoscope see Rawson (n. 5) 308 on D.S. 2.29–31.

of the client, rather than the methods of the astrologer, is under attack: it is 'human nature to prefer to believe in mysteries'.

The connections of magic, too, with respected fields of learning posed a threat. It is at the doctrines of the Magi that Pliny's criticisms are most frequently directed in the *HN*. The influence of their creed compels him to give a detailed outline of their origins and history (30. 1–18). References to their charms and spells occur frequently, especially in the books (20–32) which deal primarily with medicines.[29]

Pliny saw magic as arising from a degeneration of medicine (30. 2), adding to itself the powers of religion and then of astrology. He recognizes that there is no sphere of magic at once separable from religion and science.[30] Hence, it is an even greater threat to the truth as Pliny sees it, and he makes all possible attempts to isolate it from 'true' medicine and religion. He even places all blame for magical doctrines on one specific group of people, the Magi. This is probably not an exaggeration on Pliny's part. A. D. Nock, despite arguing that the term *magus* was generally used in a sense of disapprobation, and did not refer to a specific group (of Persian priests for example),[31] allowed that Pliny's usage is more definite than that. He suggested an identifiable body of literature, perhaps forged in reply to popular demand and the desire for completeness in the library at Alexandria.[32] Such a theory fits in very well with the virulence of Pliny's attack. Magic is not merely a case of debased superstition practised unofficially by the uneducated *vulgus*. Here is a body of literature, forged perhaps, but capable of challenging or infecting traditional learning and of tempting a more intellectual audience.

[29] *HN* 30. 1–17. *Vanitas, commenta, monstra Magorum*: 22. 20; 24. 156–67; 25. 13, 106; 26. 18, 20; 28. 47, 85–6, 92 ff., 228–9; 29. 53, 68, 76, 81; 30. 18. This list is by no means exhaustive. There are also many Magian remedies in bks. 20–32 which Pliny inserts without comment.

[30] A. D. Nock, 'Paul and the Magus' (n. 18), 313–15.

[31] Nock (n. 18) 321, cf. Rawson (n. 5), 309.

[32] Nock (n. 18) 322 ff. He picks out the credulous Greek historian Hermippus, who Pliny says was a writer on every aspect of magic and an exponent of two million verses by Zoroaster! Such books may have been based on Persian ideas, even books, but were nowhere near actual translations. Cf. P. Green, 'Prolegomena to the Study of Magic and Superstition in the Natural History of Pliny the Elder, with Special Reference to Book xxx and its Sources' (diss., Cambridge, 1952), 157–8 on Hermippus; he too makes the suggestion that the object of Pliny's attack is a learned magic, a *doctrina* associated with the Eastern 'Magi' (111–18).

Pliny sees a dangerous ambiguity in magic's origins and pretensions. In his most forthright condemnation, he complains that it has *umbrae veritatis* (30. 17) but that these are the arts of the poisoner, not magic. The term *veneficium* is ambiguous (see below, sect. 2. A. iii). For the moment, we need note only the significance of *umbrae veritatis*: magical learning is deceptive. It is devoid of its claimed supernatural powers and is merely a perversion of a respectable *ars*, medicine, into its opposite, poison. It remains to examine how each part of this deception threatens ideas basic to Pliny's world view.

(ii) *Manipulations of supernatural power: magic and* vis naturae

The connection formed by the concept of *vis* between magic and Pliny's *Natura* has already been mentioned (sect. 1). The magician believes that *vis* is there to be manipulated. Pliny continually denounces his attempts as frauds and failures. It is all *vanitas* (30. 1; 30. 19), *fraudulentissima artium* (30. 1) and *intestabilis*, *inrita*, *inanis* (30. 17). Pliny is, of course, a sceptic with regard to conventional beliefs in the supernatural. He cannot believe that words in ritual prayer-formulae and magical incantations can have any effect, since he does not believe in the spiritual beings thus invoked. For him, traditional belief in an afterlife is also *vanitas* (7. 188). Reported instances of ghosts are an awkward intrusion into his view of Nature, dismissed as *prodigia*, not *naturae opera* (7. 179). Nero was eager to call up ghosts by magic, but failed. His lack of success is proof of magic's *vanitas* for Pliny. He throws in for good measure remarks on foreign origins and a humanitarian one on the *saevitas* which created more ghosts by temporal authority than could be called up by supernatural means (30. 15; see III, intro.). If Pliny is less harsh towards the spells and charms of rural Italy, his scepticism is evident. He is too embarrassed to repeat charms (*verecundia*, 28. 29) or else he expresses his scepticism by ironic use of *mira*, as in 27. 107 on a plant which drives away what are 'mira persuasione' called *Fatui* and which he rationalizes as hallucinations. In *HN* 17. 267, he is torn between respect for tradition—Cato records such charms—and his belief in their irrationality: 'so wondrous (*mira*) are some of the devices invented by human skill, inasmuch as most people believe that hailstorms can be averted by means of a charm, the words of which I should not for my part dare seriously to put in my book'.

In one sense, then, Pliny's cries of *vanitas* are an impatient dismissal of mankind's follies, tinged with anxiety over the fear created by these as by other unfounded beliefs. Often, too, Pliny suggests that the Magi build up a wholly spurious aura of power around their remedies by choosing unnecessarily bizarre or disgusting ingredients. To the modern reader, there is often little to choose in this respect between Pliny's magical and his 'respectable' herbal remedies. However, on the rare occasions when Pliny himself is noticeably startled, it is usually by a magical remedy: 'Who could give to one delirious the brain of a mouse . . . or the ash of a weasel, or even the dried flesh of a hedgehog, even if the treatment were bound to be successful?' (30. 95). In a Magian amulet against fever, 'why was an owl's claw chosen rather than anything else?' (28. 228–9).

Yet the vehemence of the denunciations suggests that this is not the whole story. In fact, the magicians' claim to master *vis naturae* violates the whole basis of the man–nature relationship in the *HN*. Pliny is, as we have seen, sceptical when man exceeds his inferior *ratio* in overambitious attempts to investigate nature and her properties; whether on the cosmic scale of the astronomers, or on the microscopic scale of the drug-mixers (Ch. I. 4). Clearly, then, the more ambitious the magician's claims to control the powers of nature, the less likely he is to succeed. In fact, the magician makes complete nonsense of his claims because he considers that his spoken or written charms and spells harness more power than is plausibly inherent in the natural ingredients, such as plants, which he might use (26. 18. ff.). Ointment from the eyes of a python may help quell panic caused by nightmares, but can an amulet consisting of the snake's head, together with lion's hair and marrow, the foam of a victorious racehorse, and the claw of a dog, really make a man invincible? Pliny clearly thinks not (29. 67–8). Elsewhere, he comments drily that, if a salamander really put out fire, the all too flammable city of Rome would have found out by now (29. 76).

Furthermore, *Natura* is divine, and man's investigations are, in Pliny's view, often tinged with irreverence. In this respect, the magician violates the man–nature relationship regardless of whether he actually achieves control of natural forces. His very intention is unacceptable. Lack of reverence is a hallmark of the magician. As he commands rather than supplicates the supernatural powers, so too he tries to master for his own use what should be investigated only for the purposes of greater understanding and admiration of *divina Natura*. Pliny, when describing Nero's interest in magic (*HN* 30.14), says that the emperor's greatest wish was 'to give orders to the gods'.

On a less ambitious scale, the magician has a better chance of harnessing some natural power. If his herbal potions cannot draw down the moon, they may yet have the potency to kill, madden, or maim other living beings. Pliny sees man as inflicting more localized control over parts of *divina Natura* by the misuse and perversion of her natural materials. Similarly, magical meddling can misuse the material manifestations of *vis naturae*; power that is both natural and divine. It is significant that Pliny mentions abortives and love-potions in the same list as *alia magica portenta* (25.25, see below, sect. 2. A. iii, cf. Ch. VI. 1. C.), which suggests that he considered them to be in the same category. He sees things which pervert natural patterns of life as bearing a relationship to magic. Also included in this list are drugs 'qua mentes solverentur'. Drugs which cause madness could be said to pervert *ratio* itself. It is possible that those *herbae magicae* listed by Pliny in (24.156–67) which are not definitely fictitious were hallucinogenic plants of eastern origin.[33]

For Pliny, the grandiose claims and the supernatural element popularly attributed to magic are nonsense. In so far as it involves the misuse of *vis naturae* inherent in plants, however, magic is a material danger. As we have seen, he regarded magic as having sprung from a perversion of religion and medicine. It is with regard to medicine, the element which relates to the practicalities of *vita* for man, that Pliny is at his most polemical against magic. It is in medicine that the magician creates his *umbrae veritatis*.

(iii) *'Arts of the poisoner': magic opposed to* humanitas

The arts of the poisoner aim to harm men. The poisoner is the antithesis of Pliny's ideal, a man who deliberately seeks out those

[33] See J. Stannard, 'Medicinal Plants and Folk Remedies in Pliny, *HN*', *HPLS* 4 (1982), 19, quoting H. Brugsch, 'Die Alraune als altägyptische Zauberpflanze', *Ztschr. f. ägypt. Sprache* 29 (1891), 32.

substances which *Natura* in her benevolence has generally tried to hide from or render unattractive to men (*HN* 33. 3). He is therefore against nature as a whole as well as man in particular. A connection between magic and poison is illustrated by the use of the word *veneficium* to denote magic as well as poison. A general ignorance of the nature and working of magic made it impossible to distinguish between natural and unnatural causes and the term *physikos* was eventually to get a second meaning of magic.[34] More surprising, in view of this impossibility of dividing magic from religion or science, is Pliny's conviction that the art of the Magi was harmful. After all, they advertised *remedia* as well as poisons and curses. It is true that Latin usage displays a greater awareness of the criminal than of the curative powers of magic. What is suggested specifically by the term *veneficium* is echoed in a more general way in another synonym, *ars nefaria*. Moreover, anyone with a vague awareness of the traditional Persian associations of magic would tend to emphasize its more sinister side, cf. Plut. *De Iside* 46 (369 E–F). Again, if, as has been suggested, Pliny's knowledge of the Magi was from a distinct body of writing, it was already conveniently separated from those Roman religious practices which were uncomfortably similar. His patriotic bias also aided the separation. Whether specifically Persian (30. 3) or from a number of countries (25. 13), magic was foreign and as such suspect.[35]

But most importantly for Pliny, the prescriptions of the Magi threaten his humanitarian principles because they deny man's well-being in the widest sense.[36] In *HN* 25. 24–5, Pliny proclaimed his intention of not including in his work any potions

[34] The natural and the unnatural: see in general G. E. R. Lloyd, *Magic, Reason and Experience* (Cambridge, 1979). *Physikos*: see K. F. Smith in J. Hastings (ed.), *Encyclopaedia of Religion and Ethics* (Edinburgh, 1908–21), viii. 276. The terms *venenum* and *medicamentum* were used of both 'natural' and 'magical' preparations. (A further ambiguity was that they could be used either to kill or cure: cf. Garosi (n. 10), 45–7).

[35] As was medicine, other than the native herbal remedies. The Greeks and their medical theories are under constant attack in the *HN*. See Ch. VI, especially sect. 2, intro., B–C; cf. also the kill-or-cure ambiguity of *medicamentum* above, n. 34.

[36] North (n. 10) also mentions this as Pliny's main concern, as does Green (n. 32). This is not to deny Pliny's tendency to link the uncivilized nature of certain magical practices with their foreign origins. He criticizes Greeks for adopting remedies from human bodies that originated from Osthanes the Magus and *barbari externique* (28. 6–7, cf. 30. 8, Osthanes taught magic to the Greeks). He admits the universality of magic, its inclusion in the Twelve Tables, and the

which might cause physical harm or death. Certainly, he does not relate any Magian poisons which he believes might actually cause the harm intended by their promoters. An interesting exception occurs in *HN* 28. 86, where the Magi misuse a medical method rather than a herb. It is included because Pliny is in two minds as to whether it could really work or not. In either case it is to be condemned: 'I will give the *most moderate* of (the Magi's) promises: take the parings of a patient's fingernails and toenails, mix with wax, say that a cure is sought for . . . fever and fasten them before sunrise on another man's door as a cure . . . What a fraud (*vanitas*) if they lie! What wickedness (*noxia*) if they pass the disease on!'[37]

The Magi may recommend *remedia*, but in this case disregard the well-being of a third person. In other cases, their cures involve inflicting a deeper injury on mankind by disregarding the dignity of man. In *HN* 28. 4 ff. the Magus Osthanes is apostrophized as 'eversor iuris humani': by taking remedies from human bodies, he has made *veneficia* more innocent than *remedia*. Such actions lower men to the level of beasts. The context here is that of medicine rather than of magic, but the distinction is, as we have seen, irrelevant. The humanitarian principles espoused by Pliny, which he believes should motivate the proper practice of medicine, are ignored by the Magi. There is a religious ring, too, to Pliny's language of condemnation. His criterion, he says, is to pick out *auxilia*, not *piacula*. The latter word implies more than the material harm of *veneficia*; the sancitity of Nature is under attack through such outrage on humanity.

(iv) *Magic and medicine: drawing a distinction*

Magic, then, is to be regarded with suspicion, even if it claims to effect cures or other things beneficial to man.[38] Irrational trust in

practice of human sacrifice in early Rome (30. 12). Yet he emphasizes the debt owed to Rome for abolishing human sacrifice among other nations, in particular the Gauls. Garosi (n. 10), 58–9, noted the fusion for the first time in this passage of two elements in the growing awareness of magic in the later Republic: the Greek tradition about the Persian Magi combines with growing Roman knowledge of Gallic customs as recorded, for instance, by Caesar (*BG* 6. 16).

[37] Pliny regards transference of a disease from one subject to another as an acknowledged medical method; cf. *HN* 30. 64, an apparently non-magical cure for colic made by placing puppies on the patient's stomach, and letting them suck milk from his mouth. Presumably, the practice was acceptable to him if the third party was not human.

[38] Magian remedies for fever are included reluctantly (30. 98) because there are

the supernatural powers overrules rational assessment of the natural potential of the ingredients used. The bizarre choice of ingredients (30. 95), not to mention the gibberish of the incantations (28. 20), can offend Pliny's practicality. Ingredients can also offend his sense of the dignity of man (above, sect. 2. A.iii). These doubts produce the cumulative feeling that the Magi have a perverted view of Nature. To Pliny, it is unique evidence of their *vanitas* that they look with greatest awe upon, and attribute the greatest powers to, moles, creatures who are cursed by nature in their blindness and their living entombment (30. 19). Other animals with 'unnatural' habits are similarly revered. The cricket, which walks backwards, bores into the ground, and chirrups at night, is assigned great *auctoritas* by the Magi (29. 138). The tick, 'most disgusting of animals', is held in high regard 'on the grounds that it has no outlet for its gorging nor any end of it other than at death, living longer if it starves' (30. 82). The hyena provides Pliny with several chapters of Magian charms (28. 92 ff.). Its *monstrifica natura* included an annual sex-change (cf. 8. 105 ff.).

Despite Pliny's efforts to explain his aversion to the Magi, however, the lack of division between genuine, and fraudulent 'magical', uses of herbals remains a difficulty for him. Nowhere is this better illustrated than in his description of the history of botany in book 25. He recognizes that popular fascination with spells and charms was an important part of the history of the discovery of the medical usages of plants (25. 10–12). He then relates how the Greek writers on herbals got material from the Magi of Persia, Ethiopia, Arabia, and Egypt. So amazed, he continues, were the ancients who read the books written under Magian influence that, relying on them, they made unbelievable statements. However, although these are incredible (*fide carent*), 'yet they fill us with wonder and force us to admit that there remains some truth in them' (25. 13–15). Some authorities, he adds, claim that there is nothing that plants cannot effect, but that their powers are not yet fully known. Pliny's dilemma is clear: in dismissing the inflated claims of the Magi and their believers, he risks underestimating *vis naturae*.[39]

no others. Thérasse (n. 8), 316, does not do Pliny justice when he accuses him of contradicting himself by this inclusion.

[39] Nock, 'Graeco-Roman Beliefs', *Essays* (n. 18), i. 37 nn. 29 and 30 says that the desire for divine power—*arete*, *dynamis*, *virtus*—is especially clear in the invocation of herbs with occult powers.

There are other problems in distinguishing between right and wrong uses of herbal powers. The magician can be seen as coming between plants and their beneficiaries and distorting their use. Yet, as we shall see (Ch. VI. 2. A), the professional doctors, the *turba medicorum*, do much the same in Pliny's view. Their claims for their overelaborate theories are as nonsensical as the magicians' exaggerated claims for their supernatural powers.

In one respect, however, the lore of the Magi could be distinguished from Roman herbal medicine. Magic possessed the coherent body of literature and learned interest described earlier (sect. 2. A. i). Pliny has to acknowledge that, in this respect, Roman herbal lore is at a distinct disadvantage. In *HN* 25. 4, he says that there is a scarcity of Roman writers on the subject. Further on, he says that many herbs are not familiar because they are known only to illiterate country people. While many of the plant's properties are known, they still lack names. Moreover, those who have the knowledge are unwilling to impart it to others as if in doing so they would lose it themselves (Ch. VI. 2. C). It was not that herbal medicine was a static or dying art.[40] Pliny takes pains to record many recent discoveries in the field (Ch. II. 1). However, experimentation, if dedicated, was at best informal (25. 37), and elsewhere Pliny can comment that there is no certain method of discovery, 'ratio inventionis anceps' (25. 16, cf. Ch. II. 2). The stress on the power, and on the reluctance to define or impart it to others, does, of course, illustrate the 'magical' ideas which are inseparable from this native herbal lore (VI. 2. C). As a whole, however, this list also confirms what was suggested in Ch. II. Far from being opposed to theoretical knowledge *per se*, Pliny recognizes its importance in disseminating the practical knowledge of the benefits of Nature among all men and ensuring its continuity among their descendants.[41] Literature on herbal medicine would aid this process and Pliny probably saw his own writings as contributing to this cause (see

[40] See J. Stannard (n. 33), 3–23.

[41] G. E. R. Lloyd, *Science, Folklore and Ideology* (Cambridge, 1983), 135–49 discusses the pursuit and transmission of scientific discoveries in the context of these botanical passages in Pliny. Pliny's theoretical approval of first-hand inquiry is not adhered to by him in practice to any great extent. The fact that there is by now a considerable literary tradition to be preserved and handed down is one of the factors which forces the more literary approach on him (see Ch. II. 1 on Pliny and the transmission of learning).

Ch. VI. 2. D). It would also reveal the secrets of nature without that suggestion of exclusivity and obscurity with which the adherents of magic tried to enhance their powers. Pliny describes with enthusiasm the possession and translation into Latin of Mithridates' book on antidotes to poison. The book was obtained by Pompey after his victory over that king, and was translated by his freedman Lenaeus: 'and so that victory was no less profitable to *life* than it was to the state' (25. 5–7; see Ch. VI. 2. c).

B. PHYSIOGNOMY

From the extravagant claims of a creed which forms a major preoccupation of Pliny's, we turn to a topic on an altogether smaller scale. Magicians and astrologers are concerned with the links between man and the powers of the universe. The physiognomist seeks premonitory signs of personal character and destiny and looks no further than the physical attributes of the individual concerned. Pliny mentions physiognomy only once (11. 273–6). Yet this smaller topic is the more easily encompassed, and nowhere are important facets of Pliny's opposition to certain beliefs more clearly illustrated than by close examination of this compact passage. We see in a nutshell the infiltration of a superstition into philosophy, but also its tendency to become simplistic, forcing an artificially rigid system on a trivial basis. Such distortion of rationality causes unnecessary fear and instability among credulous men.

First, a summary of the passage. Pliny expresses surprise that Aristotle believed and published his belief, in physiognomy. Pliny thinks that the belief is unfounded and hesitates to discuss it because of the unfounded anxiety it could cause among those who looked for the signs in themselves. Nevertheless, he yields as Aristotle is so learned. There follows a list of indicative signs from Aristotle's *Historia Animalium*:[42] teeth (2. 3, 501^b22), lines on hands (1. 15, 493^b32), and (taken from Pompeius Trogus) foreheads (1. 9, 491^b11–15), eyes (1. 10, 491^b24 ff.),[43] and large ears (1. 11, 492^b3–4).

[42] Pliny probably read this at first hand. Its material turns up elsewhere in the *HN*'s books on animals, and Trogus, who is directly quoted in § 275, is not cited at all for the points in § 274.

[43] Pliny or Trogus makes a mistake about large whites showing impudence. Aristotle attributes impudence to those whose eyes are open and staring. In *HA* 1. 10, 492^a, whites are said to be much the same in all animals.

Pliny is torn between doubt and respect for the learning of Aristotle. To begin with, let us see how philosophy came to assimilate such an idea and why Pliny still has doubts.

Physiognomy may have been of oriental origin. To be accepted into philosophy, it needed the background of the Aristotelian close body–soul relationship rather than the dichotomy of Platonism.[44] The Stoics could link it with the doctrine of the emotions as corporeal, introduced by Zeno, together with the general Stoic theory of the effect of the emotions on the soul.[45] The doctrine implied a monistic concept of body and soul and may be compared with the idea of *pyr technikon*, which is corporeal yet spiritual. Secondly, Zeno's interest in logic possibly caused him to favour the science. The pseudo-Aristotelian treatise[46] *On Physiognomy* does speculate (4, 809^{a}20) on a physiognomical method which would appeal to philosophers since it was deduction. If a man is a rascal and poor, we can deduce that he will be a thief and a miser. Posidonius' interest (Sen. *Ep.* 95.65) may have been inspired by his ethnographical work; see, for example the ethnographical content of such passages as *On Physiognomy* 1, 805^{a}24–8.[47] Physiognomy thus acquired a respectable place in both these philosophical schools. Yet Pliny, despite his own Stoic views and his respect for Aristotle, is not impressed and one of his reasons is directly connected with a philosophical controversy, the Stoic position on Fate. His distrust of fatalism in astrology was mentioned earlier (sect. 1). His rather tortuous approach to the problems of omens and cosmic sympathy (2.97; see sects. 1, 2. A. i) suggests he was not happy with the idea, even when in a respectably learned context. In this, he was no doubt influenced by the writings of Cicero, for whom he had a great admiration.[48] Cicero criticizes the implications behind physiognomy in his treatise on the general problems of Fate, *De Fato*. Environment affects the character but not the

[44] See J. Schmidt, *Physiognomik*, in *RE* xx. 1. 1064–74; U. Dierauer, *Tier und Mensch im Denken der Antike* (Amsterdam, 1977), 104.

[45] Cic. *Acad. Post.* 1.39; D.L. 7.173.

[46] On the nature of this work, an amalgamation of two similar treatises, see R. Förster, *Scriptores Physiognomici*, (1893), i, intro., xviii ff.

[47] See E. C. Evans, 'A Stoic Aspect of Senecan Drama', *TAPA* 81 (1950), 169–84.

[48] Cicero is accorded a full-blown rhetorical acclamation in 7.116–17.

will, which can modify the character (7 ff.).[49] He goes on to describe the laughter of Socrates' pupils when Zopyrus read Socrates' 'bad' character in his physical features. He concludes that 'it is possible that these defects may be due to natural causes, but their eradication and entire removal . . . does not rest with natural causes, but with will, effort, and training' (11). This suggests that, at the most, Cicero admitted that natural physical characteristics might indicate character propensities, but that the latter can be remedied by will, even if the physical signs remain. Pliny's own interest in man's free will appears in his concern for how man's *ratio* can either follow or pervert *Natura*. The tailpiece to the story is therefore relevant: '(Socrates) said that these propensities had been implanted in him, but he had banished them from him by means of *ratio*' (*Tusc*. 4. 80). This story is repeated by Alexander of Aphrodisias in a passage describing the Stoic attitude to Fate. It concludes his exposition of the Peripatetic position on determinism (*De Fato* 6). Such emphasis suggests that the Peripatetics may have attacked the Stoics' position on physiognomy in connection with their controversial position on fate, at the same time justifying their own position with this story.[50]

When the Stoic system began to concentrate more upon ethics and the striving of the ordinary man towards virtue, the principles of physiognomy must have seemed problematic. Few would be without some slight physical abnormality and the majority would therefore risk being classified as morally imperfect, regardless of their personal efforts. Recognition of this problem can be discerned in Seneca's writings on the nature of the emotions. In *Ep*. 106, he sets out a logical argument for the

[49] Ethnographical character distinctions were not necessarily believed to be fixed. Polyb. 4. 21 is convinced both of the possibility and of the need for men to modify their natural characters. The Arcadians introduced music into their culture to temper the stubborness and harshness of nature. He deplores the savagery of the Cynaethians and prays that they may make themselves gentler by education and especially by music.

[50] The philosophical controversy about determinism and the particular problems raised by Alexander's treatise are discussed by R. W. Sharples, *Alexander of Aphrodisias on Fate* (London, 1983), see Ch. 1 n. 6. On this particular passage, he suggests (130, cf. 13) that the story was originally a protreptic to philosophy and the connection with determinism was later and secondary, perhaps due to Carneades.

corporeality of virtue, using strict formulae (106. 5 ff.). Further on, he adds: 'You will have no doubt . . . that emotions are bodily things . . . Do you think that such evident marks of body are stamped on us by anything else than body? And if emotions are corporeal, so are diseases of the spirit . . . and all the faults which harden into our souls to such an extent that they get into an incurable state' (106. 6). 'Do you not see how bravery makes the eyes flash . . . how joy produces calm?' (§ 7). 'Only a body can control or forcefully affect another body' (§ 10). In § 11, however, he voices discomfort: such logical subtleties are irrelevant to practical morality. 'Subtlety is wasted away on useless subjects. Just as we waste other things on uselessness, so too do we waste philosophy.' We suffer, he says, from excess in literature; thus we learn our lessons not for life but for the lecture-room. This is not just a dismissal of too much theory. In point of fact, Nature (*Ep.* 66. 4) 'produces certain men who, though hampered in their bodies, none the less break through the obstruction . . . the soul is not disfigured by the ugliness of the body, but rather the opposite, the body is beautified by the comeliness of the soul'.[51]

Seneca's conclusion here brings us to another facet of Pliny's doubts. The science of physiognomy, like those arguments which support it in Stoic philosophy, is too arbitrary and artificial to be of practical philosophical use. The most it may do is to tell of a theoretical propensity in the case of permanent physical characteristics, as with Socrates. It says nothing of the actual aspirations of the person within. As for the fleeting changes in looks (mentioned in *Ep.* 106. 7), who is to tell whether they are signs of a brief bout of an unwonted emotion, or are in fact a manifestation of a deeper-seated disease of the spirit?[52] It is precisely this danger, of rigid diagnosis, which Pliny fears in Aristotle's bald equations: 'Yet he does not, I imagine, note all these attributes present in one person, but separately, trifling things in my opinion . . .' (11. 274). Important predictions are being based on trivialities, a complaint which, as we shall see in Ch. IV. 3. A,

[51] This also exposes the difficulty of taking literally the Stoic idea, attributed to Zeno and Chrysippus, that the wise man will feel affection for the young men whose beauty shows their propensity to virtue, and that beauty is the flower of virtue (D.L. 7. 129–30).

[52] In *Ep.* 75. 11–12, Seneca discusses the difference between deep-rooted *morbi* and *adfectus*. Can physiognomy tell us much if *adfectus* can from time to time attack even those *perfecto proximi* (75. 12)?

he makes about prognostication from individual physical parts of animals. A similar sentiment is expressed in Cicero's *De Divinatione* (1. 118), where it seems that Stoics find prediction from bird-song or entrails 'neque decorum . . . nec dignum' partly because it is too trivial and partly because it 'restricts the divine power to so narrow a compass'.[53] This is essentially Pliny's point.

It is interesting that a later reputable 'believer' in physiognomy, Galen, used it not as a science in and for itself, but merely as a prop to medical prognostication, though his connection of it with astrology probably encouraged the superstitious form it took in medieval times.[54]

Even more revealing is the fact that even its exponents in the 'Aristotelian' treatise recognized that the bald use of physiognomical examination was not on its own a viable science. It cannot be an automatic and isolated procedure that depends for character summarization on one neat observation. Pliny's complaint is dealt with at 2, 807^a: 'It is silly to rely on a single sign'. At 1, 805^b1–10, the author says that the number of inferences which can be drawn from facial expressions alone is small. He touches on the difficulty of mechanically matching soul and body in the corrupt passage 4, 809^a10–14, when he points to the problem of differentiating the pallor of fear and that of fatigue. In fact, 'it is safest to refrain from all positive assertion when you find that your signs are inconsistent and contrary to one another in detail' (2, 807^a25). It is also interesting that the treatise is supposed to have been a combination of two separate works, since even the format of the list of characteristics and their meaning in the first treatise is more cautious and less rigid in its equations than in the 'second' treatise (6, 810^a15 ff.). In the latter, one sign is given, and the character it indicates; in the former, a list of several different signs is given to denote a particular character.

We should return briefly to Seneca and his difficulties. His interest in physiognomy has been seen in his use of portraiture in his plays.[55] We can go further and say that his interest in general lies only in its dramatic or rhetorical possibilities and that he does

[53] Pease ad loc. See also Ch. IV. 3. A and ibid., n. 31.

[54] See E. C. Evans, 'Galen the Physician as Physiognomist', *TAPA* 76 (1945), 287–98.

[55] See Evans (n. 47), 169.

not see it as having a comfortable place in his ethically based philosophy. Even in *Ep*. 95.66, where he talks of Posidonius' 'ethology', also called *characterismos*, he uses terms which have a rhetorical meaning.[56] Again, if we look closely at Seneca's enthusiastic comparison of Cato's bravery to the picture of the spirited war-horse in Virgil (*G*. 3.72ff.; see *Ep*. 95.67–71) the lack of any very specific physiognomical comparison in the Virgilian picture and the use of a piece of poetry rather than a direct physiognomical description suggests that Seneca's prime purpose is rhetorical effect. The spirit of Virgil's horse comes across as a result of a combination of all its gestures and movements as well as physical features. Seneca is talking of someone who is dead; someone, that is, whom he never saw except, perhaps, in artistic representations, and who would be known to him only by literary descriptions. The latter would, moreover, include deeds and speech, probably to a greater extent than looks. This is a very generalized use of physiognomy, relying on other background knowledge, not just a single physiognomical scanning, and with an important motive of rhetorical effect.

In conclusion, then, it is very probable that physiognomy in its crudest form—one sign designating one characteristic, however trivial the sign—became the idea which gained credence among the *vulgus*, whose credulity in the case of equally simplistic astrological equations Pliny criticized in *HN* 2.28 (above, sects. 1, 2. A. i). To emphasize his point, we have only to turn to the late second century, and Artemidorus' condemnation of other types of divination as opposed to his own dream-interpretation and see what sort of company popular physiognomy might keep: '. . . diviners by *astragaloi*, or by cheese, or by a sieve . . .' (*Oneirocriticon* 2.69).[57]

We come finally to Pliny's criticism of physiognomy on the grounds that it causes fear. We saw (Ch. II. 3) that he wished to eliminate fear, cancelling man's weakness in the face of the unknown or incorrectly known with his strength, *ratio*. Fear, is, of course, an important stimulus to those who turn to fatalistic

[56] Quint. 1.9.3. *Characterismos* and *ethologia* are equivalent to the rhetorical term *descriptio* (Cic. *De Orat*. 3.205, cf. Suet. *De Gramm*. 4).

[57] Personal animosity may have prompted Artemidorus' remarks in this passage. See R. J. White (trans. and comm.), *Oneirocritica* (New York, 1975), 155 n. 140.

creeds, believing that astrology, magic, or physiognomy will relieve this fear by certainty (above, sect. 1). Here, Pliny is worried lest people should anxiously seek ('anxie quaerat') Aristotle's *praescita vitae* in themselves.

As we saw in the first part of this chapter, such fear is often closely associated with a lack of rational control and an instability that is communal as well as individual. The obtaining of unofficial predictions of the careers of the powerful by means of astrology for instance threatened them and could cause popular disturbance. Astrological predictions of death-dates of emperors and political opponents are a recurring cause of anxiety and accusation under imperial rule. Physiognomical signs, says Pliny, are trivial, *volgo tamen narrata*. The signs are like astronomy in that they can denote length of life, and the popular belief recorded by Suetonius (*Titus* 2) that Narcissus called in a *metoposcopus* to examine Britannicus and declare he would never be emperor is a relevant case.

C. 'WEIGHING LIFE IN AN IMPARTIAL BALANCE': PLINY'S SOLUTION

In conclusion, we may return to the general denunciation of various religious ideas in *HN* 2 with which the chapter opened. In *HN* 2.25, Pliny summarizes his feelings. 'This series of instances entangles unforeseeing mortality, so that among these things only one thing is certain—that nothing certain exists and there is no creature more pitiable or more presumptuous than man.' Pliny here acknowledges the human search for certainty and security which can act as a stimulus to the beliefs he criticizes. Man would do better to recognize the impossibility of certainty in life. This does not mean that he should go to the extreme of deifying Chance itself (2.23). *Fortuna* is the personification of the untrustworthy, as Pliny's description in *HN* 2.22 shows, yet men place their trust in her alone. The result is the same passive submission as occurs in the case of the adherents of the inexorable certainties of astrology and Pliny sets them side by side (2.22–5).

The vicissitudes of man's lot, *sors* or *fortuna*, are described at some length in book 7, which is dedicated to a description of man himself. Pliny begins with the development of an intellectual commonplace, the physical shortcomings of the new-born human being: the highest species of creation is ill equipped to deal with life under the dubious auspices of *Natura* the mother/stepmother (*HN* 7. 1–5, cf. Ch. II. 3). Thereafter, he describes the amazing achievements, births, deaths, accidents, physical and mental capacities, and peculiarities of various persons. These lead him, intermittently, to comment on the *varietas fortunae* of the human race; the extraordinary differences between men's lives and even within the life of one individual. As we have seen, Pliny has dismissed all those beliefs to which men have resorted in order to withstand such vicissitudes and even to gain some control over fate by finding out what the future holds. Without such religious supports, how does Pliny expect the human race to approach life's uncertainties? Indeed, how do such uncertainties fit in with his frequent portrayal of *Natura* as the beneficent, providential mother of mankind?

In fact, certain aspects of *Natura* have much in common with the fickle goddess who plays blindly with human fate. We saw in Ch. I. 3 that the image of *natura ludens*, nature at play, could be callous and capricious. She is not always *natura provida*, selflessly supplying men's needs. In *HN* 7. 32, she creates *varietas* in the form of human freaks as an amusement for herself, *ludibria sibi*. It has been observed that some of the passages in which *varietas* is used of *Natura* show her taking the part normally played by *Fortuna*.[58] If we bear this in mind, the prominence Pliny gives to the *Natura* as mother/stepmother at the beginning of book 7 is understandable.

Man's weakness from birth is complemented near the end of book 7 (190) by the insistence that there is nothing for man after death but oblivion. The text in between, however, completes what was described in Ch. II. 3 as the paradox of man: despite an unpromising beginning and an inexorable end, man's life may be filled with the greatest achievements. Fortune, or the quirks of nature, may help or hinder the attainment of these achievements. The Stoics, according to Servius (*Aen.* 8. 334), assign birth and death to fate. Everything in between they assign to Fortune.[59] It is, once again, the influence of Stoicism above all which colours Pliny's own attitude towards the problem of life's uncertainties.

[58] W. Kroll, article 'Plinius 5', in *RE* xxi. 1. 271–439, at col. 410.

[59] The Stoics' differentiation between Fate and *Fortuna* was complex and unclear, especially as they wished to play down the importance of *Fortuna*

Hellenistic philosophy, but especially Stoicism, developed the idea that man's mind could overcome fortune.[60] For Romans, the idea coincided with their appreciation of *virtus*:[61] Cicero said *Fortuna* was the mistress of external, corporeal things (*Tusc.* 5.25) and was inferior to *ratio* (*Tusc.* 2.11). The man who has *virtus* (Sall. *Jug.* 1.1–3) has no need of fortune. Undeserved good luck and chance were opposed to the life of continual mental and physical striving after right. The concept of 'varia et libidinosa . . . domina' (Sen. *Ad Marc.* 10.6) was at variance with the Stoic doctrine of the wise man's being his own master. Stoic writers frequently express the idea that *constantia* can break the power of fortune.[62]

The victory won by such Stoic *constantia* is moral rather than material. In *HN* 7.91 ff., Pliny's examples of outstanding bravery include lists of the spoils captured by the military heroes whose exploits he recounts. But the end of his account suggests that true achievement may not be material success. 'Although these cases exhibit great achievements of *virtus*, yet they involve still greater achievements of fortune' (7.104). He passes on to the climax of the list of *exempla*, that of Sergius (above, sect. 1), who conquered bad fortune as well as men. He did not receive so many of those prizes that were taken as signs of worldly good fortune, such as wreaths and triumphs, since a number of his victories were won in military defeats. Sergius exemplifies the paradox of man for Pliny.

The wise man must develop inner resources to regard the rewards and penalties of Fortune with equal indifference.[63] In *HN* 7.43, Pliny chides the overconfident man who thinks he is 'not even fortune's nursling, but actually her offspring', a phrase which, incidentally, clearly suggests the identification of *Fortuna*

(Ferguson (n. 3), 78). Seneca distinguishes between *Fatum* and *casus* in *Ep.* 16.4–5. *Fortuna* could on occasion be identified with *Fatum* (Sen. *Ben.* 4.8.3) but otherwise, it is best to see it as having, in E. V. Arnold's phrase, no existence for Stoics in an absolute sense. It is met with in practical life from the limited point of view of the individuals concerned, cf. Aet. *Plac.* 1.29.7 (*Doxogr. Gr.* 326). See Arnold, *Roman Stoicism* (Cambridge, 1911), 209, § 231 and nn. 73, 75.

[60] *SVF* i. 449; iii. 52. See M. P. Nilsson, *Geschichte der griechischen Religion*, ii[3] (Munich 1967), 204.

[61] See e.g. I. Kajanto, 'Fortuna' in *ANRW* II. 17.1 (Berlin/New York, 1981), 502–58.

[62] e.g. Sen. *Clem.* 2.6.3; *Const. Sap.* 8.3; 19.4; and *passim*.

[63] *SVF* iii. 575; Sen. *Vit. Beat.* 15; 23.2.

with *Natura* the mother and stepmother. He should remember the unchangeable *fragilitas* of man in birth and death and accordingly 'weigh life in an impartial balance' (7.43–4). The need for a balanced, impartial attitude to life's *varietas* is particularly stressed by Pliny in *HN* 7.130 ff. Fortune has dealt indulgently with the man whom it is possible to call not unhappy. Even if things go well, there is always a fear that Fortune will grow weary and, 'once this fear is entertained, happiness has no firm foundation' (130–1). The lives of those popularly considered fortunate are shown in fact to have included misfortunes as well. A senator who survived proscription by Sulla was later put to death by Antony (134). The highly successful family of the Metelli had their misfortunes (139 ff.). Even the emperor Augustus suffered from his association in the second triumvirate, from various illnesses and from discord, plots, and deaths within his own family (147 ff.).

That true *felicitas* depended upon moral rather than material welfare was, of course, another Stoic commonplace.[64] Such an attitude clearly inspires Pliny's forthright denunciation of the claims of the dictator Sulla to the title *Felix* (7.137–8).[65] He obtained his title through civil bloodshed, exiling and slaughtering thousands of his fellow-countrymen: 'O what a false meaning to attach to the title (*prava interpretatio*)!' (7.137). It is also typical of Pliny that he should attach so much importance to the idea that *felicitas* can only come through the humanitarian principle of man helping man. Seneca expressed a similar opinion, talking of 'the surest refuge against ill fortune, the haven of mutual help' (*De Clem.* 2.5.3). In *HN* 22.12–13, Pliny couples Sulla's title of *felix* with the crown he was awarded for saving citizens' lives and says he should yield both to Sertorius. Sertorius, it could be argued, had shown *pietas* by opposing his fellow citizens' oppressor. In spite of his many successes,[66] he

[64] True *felicitas*: Cic. *Fin.* 3.26 and *SVF* iii.583–8. It was equivalent to Greek *eudaimonia*.

[65] See especially J. P. V. D. Balsdon, 'Sulla Felix', *JRS* 41 (1951), 1–10; S. Weinstock's review of Latte, ibid. 51 (1961), 206–15. Kajanto (n. 61) suggests Sulla's crimes may have encouraged those educated in Greek philosophy to attempt a more moral definition of *felicitas*. Cf. Cic. *Ep. ad Nep.*, fr. 5.

[66] Sertorius had a reputation for *felicitas* in battle. On *felicitas* in the military sphere, see H. D. Erkell, *Augustus, Felicitas, Fortuna* (Lateinische Wortstudien; Göteborg, 1952); E. Wistrand, *Felicitas Imperatoria* (Göteborg, 1987), and cf. Plut. *Sert.* 15, 20.

was eventually defeated. From the philosophical point of view, however, the ultimate worldly failure was irrelevant. It was the moral effort that counted.

It is in any case difficult to evaluate happiness in life. Pliny comments that 'Fortune often spares men as a punishment' when describing the living conditions of the Germanic tribe of the Chauci. This people was spared the apparent misfortune of Roman conquest but endure the greater misfortune of living in a barren wasteland (Ch. II. 4). The human race 'counts the number of days when it is their weight that matters' (7. 132). Man tries to assess the gifts of *Fortuna/Natura* by material means. His simplistic calculations ignore moral and spiritual dimensions (see Ch. II. 4).

It cannot be disguised that *Fortuna/Natura* distributes her favours in life without any regard for just deserts and Pliny does not attempt to do so. There is no positive argument in favour of the life of moral effort as opposed to cynical passivity other than that virtue is its own reward. It can have been persuasive only to a select few. Pliny has recourse to a generalized rugged optimism. There must be a pragmatic compromise in facing life. The ambiguity of *Natura* as mother or stepmother is to be accepted but regarded optimistically. Overall, she gives more help than hindrance in life (18. 5). What is remarkable about the discussions in *HN* 7 is not so much the moral reasoning as the general tone of enthusiasm. All the facts about man recorded by Pliny here, even the twists and turns in his fortunes, are *mirabilia*, causes for wonder rather than depression. This celebration of what has been achieved by so many despite the vagaries of fortune carries a persuasive force beyond that of any argument.

IV

Man and the Animals

Introduction

So far, the discussion of Pliny's thought has centred on man and the powers that are greater than man. Man has been portrayed contemplating the divine cosmos in approved Stoic fashion, 'leaving the ground and fixing the mind's eye upon questions about the nature of God and . . . the principle which governs the universe' (Sen. *De Brev. Vit.* 19.1–2). The uniqueness of the conception of the *Natural History* becomes more apparent only when Pliny turns his attention back to earth, to the forms of life generally considered to be lower than man. Even in the *Natural Questions*, the work nearest in scope and subject-matter to the *HN*, Seneca, who says at the beginning that study of terrestrial creation is one-third of the study of the universe (*NQ* 2.1), discusses only large-scale cosmic phenomena and their relation to man. Only once is any mention made of the relationship between man and the animals and that is a gloomy prediction of the great flood, 'the extinction of both man and the wild beasts whose minds and manners man will by this time share' (*NQ* 3.30.7). The transition of man into beast reflects the extremity of the situation. Seneca normally emphasized the differences between men and animals rather than the similarities. A gulf between man and beast is characteristic of the concentration in Late Stoic thought upon man and his aspirations. This emphasis in turn took to extremes a teleological bias, an idea that the parts of nature subserved some purpose, which is recognizable in earlier ancient thought on animals: man is supreme; the animals and the rest of nature exist solely for his convenience.[1]

[1] This is not to disregard opposition to and criticism of such a standpoint. Epicureans and Sceptics did not adhere to it. In the 2nd c., Celsus attacked both Stoics and Christians for their view of man as the centre of the universe (Origen, *Contra Celsum* 4.74–99).

1. *Teleology: Beasts as Tools and Servants of Man?*

A. TELEOLOGY IN ANCIENT THOUGHT ON ANIMALS

It is worth digressing briefly, to set the background to Pliny's books on animals. Firstly, teleological ideas are to be found even in Aristotle's zoology, generally considered to be the most scientific of ancient writings on the subject. Overall, biology from its Greek beginnings was dominated by interest in man. Scientific medicine began much earlier than veterinary science. The Hellenistic veterinary treatises concentrate on an animal of paramount importance to man, the horse, which was used in war and was a symbol of power and status.[2]

Secondly, the greater importance attached to farming and the consequent familiarity with the domestic animals which provided for a large percentage of man's needs encouraged a very utilitarian outlook. Animals were looked upon as a source of livelihood and profit. It was for this reason rather than for their own sake that they were treated well: nothing pays better than taking good care of your oxen, or so we are assured by the hard-headed Cato (*De Agr.* 54. 5).

Thirdly, aesthetic appreciation is rarely of the animal in and for itself. Certainly, there may be general appreciation of animals in a natural landscape, as in Euripides' *Bacchae* for example. But when Aristotle talks about beauty in animals (*PA* 1. 5, 645^{a}23 ff.) he adds that by this he means that nature's creations show purpose not accident. Even admiration of animals in art is often admiration of the Art (of Nature) which fashioned them. Animal pictures might be classed as a humbler genre (*HN* 35. 112), and Pliny's chapters on art lead to uncertainty as to whether the

[2] Material, mainly late, may be found in E. Oder and C. Hoppe (edd.), *Corpus Hippiatricorum Graecorum* (Leipzig, 1924–7, repr. 1971). See J. M. C. Toynbee, *Animals in Roman Life and Art* (London, 1973), App., Ch. I, 303 ff. Status: an important reason for the keeping of royal menageries, for example in Ptolemaic Egypt. Beauty and novelty would arouse interest and so add indirectly to the status value of an animal. In Rome, status is one of the motives behind the stocking of *vivaria* and aviaries by the Roman nobility and the keeping of fierce pets (G. Jennison, *Animals for Show and Pleasure in Ancient Rome* (Manchester, 1937) 132).

response was emotional or a more impersonal admiration for the technical skill of the artist.[3]

In all three categories, there is a bias towards appreciation of animals for their purpose rather than for themselves. Their beauty is admired, but often because its perfection bears witness to the perfect suitability for man's use of the great masterpiece of Nature, the universe. Paintings of animals may be admired for the human *ars* which can rival that of Nature by reproducing her images so truly.

However, the existence of more complex and emotional responses to animals in the ancient world is also evident, even in practical treatises on agriculture. The lack of sentiment expressed by Cato extended to human servants, slaves of course, as well as to animals: 'see to it that your oxen are cared for with the utmost attention. Be reasonably indulgent towards your ploughmen, so that they will be more willing to look after the oxen' (*De Agr.* 5.6). Varro, on the other hand, is well aware of the demand for animals for pleasure as well as for profit: his own aviary serves as a pleasant relaxation (*De Re Rust.* 3.5.9 ff.).[4] Columella's comments on bantams, which yield little profit 'unless one takes pleasure in their small size' (*De Re Rust.* 8.2.14), and on doves,

[3] A humbler genre? In 35.112, asses are classed among 'humbler' subjects together with shop-tableaux, articles of food, etc. Otherwise the evidence is from silence. There are very few mentions of animal paintings among the works of artists which Pliny lists. The mention of Nicias' animal portraits is exceptional. Mentions of animal bronzes and sculptures are also relatively scarce: e.g. a dog is attributed to one Simon (34.90) and mention is made of Myron's ox and of works by Cephisodorus and Callicratides (34.57, 75; 36.43). See n. 6, however. *Realism* as a criterion for animal paintings and sculptures is seen in 35.95: Apelles painted a horse. When it was shown to real horses they neighed. *Technical skill*: Callicratides' marble miniatures of ants have feet so small that they cannot be discerned (36.43). Cephisodorus made, as part of a statue of Apollo, a stag so lightly poised that a thread could be passed under its feet (34.75). On Pliny's art criticism see e.g. K. Jex-Blake and E. Sellars, *The Elder Pliny's Chapters on the History of Art* (Oxford, 1896); more recently A. Michel, 'L'esthétique de Pline l'Ancien', *Helm.* 38 (1987), 55–67; and now J. Isager, *Pliny on Art and Society: The Elder Pliny's Chapters on the History of Art* (London, 1991).

[4] Varro's comments often suggest a mixture of motives—combinations of pleasure and profit and of different types of pleasure. The development of sea-water ponds 'appeals to the eye more than the purse' (*De Re Rust.* 3.17.2). Peacocks are recommended for the aviary which is built for pleasure (3.6.1), which suggests once again the visual pleasure of the striking and beautiful. Lucullus' dining-room is a combination of both pleasure and profit (3.4.3). The pleasure must have been both visual and culinary, but he had not bargained for the offensive smell. No doubt it was also a striking status symbol.

whose owners pay large sums for the sake of the pleasure derived from their ownership, indicate the existence of the sentiments which we connect with the possession of pets, *deliciae* (*De Re Rust.* 8. 8. 10). The sparrow immortalized by the poet Catullus is merely the best-known of a large number of pets commemorated in ancient literature and art. Vegetius' later veterinary treatise, *Mulomedicina*, gives the economic value of horses as justification for not scorning veterinary advice. Yet he also mentions the love and affection of their masters (*Mulom.*, prol. 11). Seneca, despite his philosophical orientation towards the supreme importance of man, is aware of the fondness for pet animals (*Ad Marc.* 12. 2: the rearing of pets is a pleasure in itself). Realism does not lead to lack of feeling or emotional responses to animals: the two are admirably blended in Virgil's *Georgics*.[5] Whatever the artistic value attached to animal representations, their number bears witness to their general popularity.[6] Finally, their variety—animals on their own, in a natural landscape, as man's servants, as his pets, and as victims in gruesome scenes of slaughter in the arena—mirrors the complexity of the ancient response to them.

What response would Pliny's five books devoted to animals have received from a contemporary audience? It is often said that

[5] Virgil's range of attitudes to animals: examples include (i) realities of farming (*G.* 3. 515, horse in a plague); (ii) recognition of human traits in animals (*G.* 1. 412 ff., ravens; 3. 220–41; 515–19, the bull-fight and the pathos of the ox mourning for its partner; 4. 3–4, the organization of bees); (iii) factual observation (*G.* 3. 417–24, a snake moving); (iv) aesthetic interest (*Ecl.* 2. 40–1, markings of animals; *Ecl.* 6. 53, white bull at rest). In *G.* 2. 209 ff. the 'ancient homes' of the birds are destroyed when trees are felled, but the author does not indulge in pity. There is a 'contrast between the homeless birds and the rich promise of plenty' of the new-ploughed land (T. E. Page (ed.), *Bucolics and Georgics* (London, 1974), 262).

[6] Pliny indicates that 'humbler' pictures in general were very popular. Piraeicus' subjects (35. 112) gave very great *voluptas* and his small pictures fetched higher prices than the largest works of art of many other artists. On the popularity of animal paintings and sculpture, see Toynbee (n. 2), *passim*. Pliny's personal prejudices may obscure the actual fame and popularity of animal representations. Thus Myron's famous heifer is not described: it is said only to be celebrated in some well-known sets of verses, 'inasmuch as most men owe their reputation to someone else's talent rather than their own'. In 34. 89 he recounts the story of Perillus' bronze bull, designed to bellow if a man were placed inside and a fire lit underneath. The story calls for an apposite outburst of indignation and Pliny obliges with a rhetorical flourish. However, his comment that 'such were the depths to which he had diverted this most humane of arts from the sculptures of gods and men' may still be taken seriously and suggest his personal evaluation of the purpose of art generally. See now Isager (n. 3), 103.

an interest in animals as pets and the like was fostered by the increasing importance of home life as political scope decreased in Hellenistic Greece and in Rome under the principate. Similarly, the idea that the conquests and exploration of Alexander played a part in introducing new and exotic animals to his subjects is worth consideration. Alexander appears in a significant number of animal anecdotes in Pliny:[7] the results of his expeditions and their finds may well have provided material for that basic human fascination for the strange and the new, giving it a fresh impetus which lasted into the following age of Rome. Aristotle bears witness to such fascination: 'in all animals there is something wonderful' (*PA* 1.5, 645^{a}17 ff.). In one passage, Pliny suggests that an animal's exotic provenance aroused more interest than its qualities and accomplishments. A certain type of magpie, he says, is less celebrated because it does not come from a long distance, but it talks more articulately than others (10. 118). Even the rather dull lists of animal-shows, indicating the number and type of animals presented, testify to this same fascination. Strange and exotic animals and birds were often brought to Rome purely for exhibition, in the Forum or elsewhere. To the emperor Claudius was sent that epitome of the marvellous, the phoenix. Although Pliny claims it was a patent forgery, it was considered worthy of exhibition in the *Comitium* (10. 5). Pliny's own emphasis on *mirabilia* in the *HN* is, as we saw earlier (Intro., pp. 8–11), a characteristic of the age in which he lived. In the early 80s, a similar outlook with a particular emphasis on animals found expression in Plutarch's *De Sollertia Animalium*. Among the most popular of animal wonders were the anecdotes about the intelligent dolphin's remarkable associations with

[7] Alexander and animals: 8. 44, 119, 149, 154; 9. 27; 10. 155. Pliny's story that Alexander commissioned Aristotle (8.44) to write his *History of Animals*, providing him with informants from every part of his empire, has been dismissed because his researches could mostly have been done in the area of Lesbos (H. D. P. Lee, 'Place-Names and the Date of Aristotle's Biological Works', *PCPS* 179 (1946–7), 7–9; cf. R. Lane Fox, *Alexander the Great* (London, 1973), 53, 509). We cannot tell how scientific Alexander's personal interest was. Pliny (*HN* 8. 119) tells how the advanced age of some deer was recognized because they were still wearing gold chains put around their necks by Alexander. There is nothing to tell us of his motive for so doing, however. Personal affection is attested by the story of Bucephalus (8. 154). Probably connotations of status, as in gifts of animals to rulers, are most relevant to him cf. the story of the hounds in 8. 149–50 (see sect. 3. B).

human beings. These were commonplace by Pliny's time and while some of those which he recounts (9.24 ff.) go back to the Greek legend of Arion (9.28), others are from relatively recent times and, it seems, were recorded by a remarkably large number of authors: a tale from Augustus' reign was to be found in 'Maecenas and Fabianus and Flavius Alfius and many others' (9.25).[8]

Recognition of this interest in marvels serves as a reminder that 'ancient attitudes to animals' cannot be summarized by concentrating on two philosophical stances, one for and one against an idea of the equality of man and beast. The crowds in the amphitheatre wondered at the cleverness of performing elephants or felt an affinity with others who begged their killers for mercy (8.19–22, see sect. 3. B). They were impressed at the apparent similarity between man and beast, but need not have taken the idea to any conclusion.

Bearing all these points in mind, we may consider Pliny's attitude to animals. He is too often dismissed as another collector of anecdotal material on animal *mirabilia*, related to the tradition of paradoxography and to the later moralizing stories of Aelian.[9] More attention should be paid to the fact that it was, nevertheless, Aristotle whom Pliny admired and to whom he acknowledged a great debt in *HN* 8.44 (see sect. 1. ii, *init.*) It is easy enough to say, on the evidence already discussed in previous chapters, that his views are teleological and that his passages on animals will reflect that. The picture is complicated by claims that Pliny's animal stories were of prime importance in influencing later writers who assigned to animals intellectual or moral qualities equal to or greater than man's ('animalitarianism'[10]). Are there conflicting attitudes in Pliny? Does he portray animals as having as much as, or even more intelligence than, man? Do

[8] *Mirabilia* and paradoxographical writing, see Intro., pp. 8–11. G. Williams, *Change and Decline: Roman Literature in the Early Empire* (Berkeley, Calif., 1978), 190–2. On dolphins, cf. n. 26. The erudite Claudius' interest in animals seems to have been more than an expression of his status as ruler. A number of unusual animals and animal feats are recorded in his reign by Pliny. Besides the phoenix, he was also sent 'a hippocentaur, carefully preserved in honey' (7.35).

[9] G. E. R. Lloyd, *Science, Folklore and Ideology* (Cambridge, 1983), 56, thus dismisses Pliny. On Aelian see sect. 2. ii and n. 23.

[10] A cumbersome word apparently coined by A. O. Lovejoy and G. Boas, *Primitivism and Related Ideas in Antiquity* (Baltimore, 1935), and used in this

animals share man's emotional responses? Do they possess moral qualities which set him an example? In considering these questions we need firstly to evaluate the extent to which his ideas are teleological and then to consider whether there is any reason to think that his passages on animal intelligence etc. are 'animalitarian'. Finally, we must consider the conclusions in the wider context of the *Natura* theme.

B. TELEOLOGY AND ITS LIMITS IN PLINY

Aristotle, as we have said, is regarded as the only ancient exponent of scientific zoology.[11] Even he views it in the light of philosophy. All things have an 'end' or 'object' (*PA* 1. 1, $641^{b}12$). There is a division between the eternal things of the Upper Cosmos and the perishable things, composed of the four elements, which inhabit the sublunary world (*Met.* 11, $1069^{a}30$ ff.). These perishable products of nature, plants and animals, are themselves arranged in a hierarchy of perfection (e.g. *GA* 2. 1, $732^{b}29$–30). However, they have the advantage of proximity for study over the eternal objects of philosophy. They are nearer to us and more akin to our nature (*PA* 1, 5, $645^{a}3$). Thus, ancient interest in man as the most perfect being need not entail emphasizing links with the divine and ignoring those with the lower forms of life. Man's physical as well as spiritual nature is worthy of study. Aristotle shows interest in the lowest forms of life (*PA* 1. 5, $645^{a}5$–25) and stresses the ubiquity of beauty in all animals: every part of the universe has its place and function in the whole, even if we are to admire it for this reason and not for its own sake.

In this sense, Pliny may be regarded as a follower of Aristotle to whom he acknowledges a debt (8. 44).[12] Lacking the latter's

chapter for want of any more convenient term. These authors claim Pliny was 'the one Latin writer . . . whose influence in favour of animalitarianism was greatest'. Those who consulted the *HN* would find the evidence to suit their argument, cf. the same authors' comments on the use of Aristotle for arguments both for and against 'animalitarianism' (390 f.).

[11] On Aristotle see especially Lloyd (n. 9), 7–57. He shows that, despite Aristotle's new 'scientific' approach to zoology, he was still influenced by popular tradition on animals. There is in his work an idea of the hierarchy of perfection with man at the top (35, 37, 42).

[12] General assessments of Pliny as zoologist may be found in L. Bodson, 'La zoologie romaine d'après la *N.H.* de Pline', *Helm.* 37 (1986), 107–16; F. Capponi, 'Cultura scientifico-naturalistica di Plinio', *ibid.*, 131–46.

degree of scientific and philosophical expertise, he nevertheless focuses on man in his natural setting and does not try to stress his different plane at the expense of the rest of nature. This attitude is, of course, to be explained partly by his pantheistic view of nature. However, when he says that the small and mean should not be despised (11. 4), we must remember that this was not the Romans' usual view. They tended instead to measure an object's interest and value in terms of size and grandeur.[13] Despite his teleology, we come quite close to a disinterested appreciation of animals in Pliny. Two basic elements of Pliny's view of the world as seen in the earlier chapters are also relevant: his practical, earth-bound view of life and his positive and optimistic approach. For him, man is exalted all the more if he praises rather than disparages the rest of the natural world; his position at the top of the scale is all the greater if the creatures on the lower steps are fully appreciated.

Pliny's overall theme is Nature and his praise of Nature through all her works also owes much to Aristotle. Perfection and artistry in animals redound to the glory of *Natura artifex* (11. 1) and to the other aspects of her power, her *varietas* and her production of *mirabilia* and *miracula*. In later Stoic thought, however, the tendency towards teleological explanations is emphasized to the point where the power and supremacy of Nature seem almost to be rivalled by that of man. One aspect of this train of thought, the idea of man as civilizer of the natural world, is particularly relevant to Pliny. As we have seen, he wrote of the harmony achieved between man and nature in gardens, where nature is under the constriction of cultivation. Her wildness is limited to the acceptable form of luxuriance in products basic to man's survival. Harmful plants are kept out: so are harmful animals.

The best expression of the Stoic idea is in Cicero's *De Natura Deorum*, 2. 99. 'Think of the human race, which has been appointed, as it were, to be the gardeners of the earth, who will not permit it to become a savage haunt of monstrous beasts or a wilderness of thorny scrub. Under their hands, the lands . . . shine out, decked with their buildings and their cities.' The Stoic

[13] Judging works of art by size: Pliny, *Ep.* 1. 20. 5; Stat. *Silvae* 1. 1. 1. On Roman taste or lack of it in general see J. V. P. D. Balsdon, *Romans and Aliens* (London, 1979), 177.

belief that animals lacked reason could be used as an argument to prove that they and the rest of creation are intended to serve 'those living creatures who are endowed with reason. These are gods and men' (*ND* 2.133). Everything else, including animals, can be tidied away by the gardeners of the universe if it is not sufficiently subservient to men's needs. As 'the Good will never exist in that which does not possess reason' (Sen. *Ep*. 124.8–13, cf. *Ep*. 76.9–10) there could be no injustice in killing animals. Certainly, there was no problem with regard to those creatures which were fierce towards man. As Autobulus, a speaker in Plutarch's *De Sollertia Animalium* (964 F ff.) says, 'there is no injustice, surely, in punishing and slaying animals which are antisocial and harmful, while taming those which are gentle and friendly to man and making them our helpers in the tasks for which they are severally fitted by nature'. The 'purpose' of some of these 'friendly' animals was to be slaughtered for food. For Cicero's Stoic Balbus (*ND* 2.160), an ox's services to man made it a crime to eat its flesh, but 'of what use is a pig, except to be eaten?'.[14]

We feel the general effect of this philosophy in Pliny's praises of Italy. Her wild animals are harmless to man while she is rich in sheep and horses which serve men's needs (37.201). This may be compared with Virgil, *G*. 2.140 ff. and with Varro, *De Re Rust.* 2.5.3, who says that the Greeks derived *Italia* from Greek *italos*, an ox, on account of the quality and quantity of the land's cattle. Strabo (2.5.33) suggests that the depredations of those who collect animals for the shows in Rome might have the effect of encouraging agriculture in places where it was otherwise made difficult by the number of destructive wild animals. In Pliny, however, the idea of man the civilizer is always balanced by the overall theme, *Natura*. Animals in all their variety, in all their physical, mental, and temperamental qualities are not mentioned only as and when it is apposite to cite their purpose in men's lives. They are described in and for themselves in so far as they are examples of the *varietas* of *Natura*. Even if Pliny's *Natura*

[14] He quotes Chrysippus: the life in a pig is merely the salt which prevents it from going bad. This view was specifically criticized by later writers attacking the idea of a man-centred universe. See K. Thomas, *Man and the Natural World* (London, 1984) on Porphyry and Thomas Hobbes (115–16, 166 and n. 2, 169 n. 26).

subserves man, she is of sufficient interest without a continual reminder of this fact. Teleology is certainly part of Pliny's view of life, but it is not allowed to restrict his aim of portraying *Natura* both in detail and as a coherent whole.

2. *Animal Qualities: Man and Beast Compared*

A. ANIMAL PSYCHOLOGY: THE QUESTION OF *RATIO* AGAIN

Pliny's interest in animals is not limited by his teleological views of nature. Yet the citation of the *HN* as an influential source book for 'animalitarian' thinkers necessitates further attention. Does Pliny ever exalt animals at the expense of man? Do his stories of animal intelligence suggest that, after all, *ratio* might not be as unique to man as we have argued? This question can only be answered if Pliny's comments are examined in the light of the ancient views on animal psychology.

A good starting-point is *HN* 27.7ff., a passage which has already featured in the discussion of Pliny's use of the terms *ratio*, *usus*, and *casus* (Ch. II. 2). It concerns the special medicinal knowledge of certain animals, exactly the type of idea which is cited in favour of an 'animalitarian' view of Pliny.[15]

'If anyone believes that such remedies could possibly have been devised by man, he shows an ungrateful attitude towards the gods' gifts . . . Surely nobody doubts that [an aconite antidote] was discovered by chance (*casu*) and that even today its every appearance is a new discovery, since the wild beasts are unable to communicate theoretical knowledge and practical application (*ratio et usus*) to each other. This is then Chance, that divinity which has made most of life's discoveries. Either supposition is equally plausible, whether we decide that the beasts rediscover these remedies from day to day or that they have an inherent instinctive knowledge (*semper scire*).'

[15] Lovejoy–Boas (n. 10), 401, cite *HN* 8. 97–101 on the medicinal knowledge of animals and mention the use by later animalitarian writers of similar stories in Aristotle, e.g. *HA* 612^a 4–10 (391).

Pliny attributes the animal discoveries to *casus*, which was defined in Ch. II as being the mind of *Natura* working through natural objects which do not themselves possess *ratio*. Animals' inability to communicate *ratio* and *usus* does not of course rule out their possession of such faculties. It is, however, at the very least, a significant limitation of their mental processes. We have seen that Pliny lays great emphasis upon transmission and propagation with regard to human knowledge (Chs. II. 1, VI. 2. D). In his opinion, the most important fact about the human voice is its use for 'expressing the thoughts, which distinguishes us from the beasts' (11. 271). Even more important are the doubts Pliny casts on animals' powers of memory, since memory is integral to the quality of the thought process. At the end of this passage, he appears to leave the question open, but a few lines earlier, he had implied a negative answer. Without memory, animals lack the ability to accumulate and preserve knowledge which is so important to man's civilization and progress in Pliny's view (Ch. II. 1).

Pliny had already implied at the beginning of the passage that human mental processes were irrelevant to the discovery of remedies by animals. The emphasis throughout on *casus* is clear. His use with regard to animals of *ratio* and *usus* in this passage is thus misleading as he implicitly denies animals a mind to any extent comparable with that of man.

Passages from other ancient authors suggest quite clearly that there was a difference between man and animal and that it was connected with rational thought. Aristotle was prepared to see only a quantitative difference between men and animals with regard to such traits of character as ferocity, mildness, fear, confidence, etc. However, with regard to human qualities of mind—practical knowledge, wisdom, and understanding (*HA* 8. 1, 588^{a}28–31)—beasts possess qualities which are only analogous, not identical.

According to Aristotle's general ethical theory, the power of moral deliberation was unique to man (*NE*. 1111^{b}4 ff., 1144^{b}24 ff.). However, he tends to give animals characteristics without making it clear that these are only 'natural' traits, not moral characteristics.[16] The distinction does become clearer when

[16] See Lloyd (n. 9), especially 25.

the Stoic equation of *virtus* with *ratio* meant that animals could not strictly share even traits of character with man. Seneca (*De Ira* 1.3.6–7) is concerned with the moral nature of man's characteristics and distinguishes between man's *affectus* and beasts' *impulsus*. Beasts, lacking *ratio*, cannot have human faults and virtues either. Similarly, Philo of Alexandria, in his *De Animalibus* (§ 96 Terian), says that animals have some vague images of respect, gratitude, and other qualities, but these are not real.[17]

It was, however, difficult to define the differences. There are some attempts to define the factors at work. An animal cannot form a train of thought; conceptual thought is not within its powers. As Seneca put it, animals are aware of their constitution but not of the definition of constitution (*Ep*. 121.10–11). Philo expresses a similar idea when he says that animals do not possess the fount of reason and thus lack the reasoning faculty (§ 96 Terian).[18]

Generally, however, it was easier to explain the differences in terms of effect. Thus Philo says that animals cannot possess rationality because this involves appreciation of abstract concepts such as understanding of God, the world, the law, national custom and citizenship, none of which animals comprehend (§ 85 Terian).

To complicate the position further, other passages suggest that even a difference based on *ratio* is no more than relative. For instance, Seneca's language may sometimes suggest that the difference between men's and animals' mental processes is only one of degree: there are animals which possess reason but only incompletely. However, he insists that there is no good in animals and hence there can be no real *ratio* (*Ep*. 124.9). Similarly, Philo insists that the imperfect reason which his opponent Lysimachus

[17] This work, a dialogue, on whether animals possess *logos*, was preserved in Armenian and later translated into Latin by Aucher. See now the edition and commentary by E. Terian (California, 1981) to whose text I refer. M. Pohlenz, *Die Stoa*^{4} (Göttingen, 1970), i.227–9 quotes other useful passages on the problem of the kinship of animal instinct to *logos*.

[18] Cf. Plutarch, *Moralia* 386 F: man alone is capable of criticism and of understanding logical consequences. See R. Renehan, 'The Greek Anthropocentric View of Man', *HSCP* 85 (1981), 247–50. That man alone had understanding rather than just perception is stated as early as Alcmaeon (24 B 1a DK).

claims for animals (§ 12 Terian, cf. § 95) is not *real* reason. In fact, the crucial difference seems to be one of development. Like animals, young children do not have *ratio* (*Ep.* 124. 8 ff., cf. Arist. *HA* $588^{a}31-^{b}3$). However, the latter gradually develop a perfect *ratio*. For Philo, too, young children possess a rational potential which animals lack (§ 96 Terian, cf. § 49). Seneca insists this is the only true *ratio* and it is peculiar to man as is implied in the language used of man in *Ep.* 76. 9–10: 'perfect *ratio* is the good peculiar to man; when *ratio* is right and perfect, man's happiness is complete'.[19]

Seneca stressed the mechanical nature of animal instinct. Animals could have memory, but it was short (*Ep.* 99. 24). Moreover, they do not acquire more as they get older. Animal instinct is not experience; all have it equally and it does not grow as life lengthens (*Ep.* 121). Animals, Seneca says, do not work through *usus* but by a natural desire for their own safety (*Ep.* 121. 20). This phrase, together with Pliny's idea of *casus* in *HN* 27. 7, is to be equated with what we call 'instinct', a mechanical sense of self-preservation. Animals do not know why they act; they have only an indistinct concept of what rouses them to action (*De Ira* 3. 7).

These, then, are the obscurities reflected in *HN* 27. 7. The whole passage suggests Pliny is unclear as to the exact nature of the animal faculties he is trying to distinguish from man's. What really concerns him, however, is not the philosophical or scientific complexities so much as the overall phenomenon as an indication of the power of *Natura*. In *HN* 8. 9–10, the instinctive fear of an animal on detecting the scent or footprint of a predator never seen before is attributed directly to 'Nature's power'. Pliny does not ask how it works; he is content to record a miracle of *Natura*. Philo, for whom God was, strictly speaking, separate from Nature, nevertheless often personifies her as the divine agency in things, talking of her 'purpose' and 'will' (§ 11 Terian and comm.). He describes the intricate behaviour of the honey-bee as being due wholly to her 'management of all' (§ 92 Terian).[20]

[19] Cf. Philo (§ 96 Terian), quoted above, who contrasts the graded division of reason in a child as it grows up with the animals which lack the capacity for such development.

[20] See Thomas (n. 14), 34 and n. 21, on Thomas Aquinas: the so-called prudence of animals was no more than divinely implanted instinct.

The part animals have to play as components of *Natura* is as important to Pliny as their function in the lives of men. What is significant is not so much the so-called superiority or inferiority of man as opposed to animal but the difference between the relation of men and of animals with *Natura*. The language of Plutarch in *De Amore Prolis* 1 is apposite because it expresses the difference between man and animals in terms of a respectively complex and simple relationship with Natura: in animals, nature is unmixed, pure and simple; in man, through reason and social life, it has become many-coloured and strange. As was seen in Ch. II, possession of *ratio* is ambiguous. Pliny suggests that the relationship with *Natura* of creatures which do not possess it is more straightforward. They are more obviously her instruments via the workings of *casus*, more obviously at one with her. In sect. 3, we shall see how animals embody for Pliny many of the ambiguities that exist in the attitude of *Natura* towards man.

B. THE 'ANIMALITARIAN' PASSAGES IN PLINY

We may now turn to the apparently 'animalitarian' passages in Pliny. First, some passages of moralization on the virtues of animals may be mentioned briefly. Here, the explanation is straightforward: the comments are incidental touches in which any implications other than a general moral point aimed at the edification of the human audience are not to be pressed. They are often used by Stoic writers. For Cicero, the swan sets an example to men since by its song it indicates that it knows death is a blessing (*Tusc.* 1. 30. 73). He is not suggesting, however, that the swan is in possession of a *ratio* superior to man's. Indeed, the swan's gift is described in vague and non-committal terms; 'they appear to have Apollo's gift of prophecy'. Such moralizations are not serious examples of 'animalitarian' beliefs. The animal is not portrayed as it really is but is idealized[21] and becomes a symbol of whatever the moralist is extolling. In Pliny, it is difficult to find any examples of this kind of moralizing with reference to individual species. The snake is said to be exceptionally faithful to its partner (8. 86) but this is mentioned as its sole saving feature. No contrast with man is ever implied. Pliny does, however, make some moral comparisons of man with beasts in gen-

[21] Lovejoy–Boas (n. 10), 19–20.

eral. In so far as these are favourable to the animals, he might be said to have 'animalitarian' feelings. Yet the very fact of comparison suggests that the original stimulus was an interest in defining and explaining man:[22] only man fights with his own species (7.5) and has unregulated appetites (10.172, 11.283). Rather than a belief in the superiority of the beasts, it could be argued that Pliny is really expressing a purely human ideal of attainment. There is a division between man and the whole of the rest of the animal kingdom implicit in such comparisons.

In comparison with a later work on animals, Aelian's *De Natura Animalium*, the *HN* has few of these superficial moralizing passages. Difference of purpose may be one reason for this. Aelian lays more stress on the artistic than on the didactic aspect of his work.[23] That the moralizing tendency should appear in Pliny's work at all is to be explained not only by the Stoic precedents but also by the desire to introduce variety of tone into his narrative.[24] In general, though, it may be noted that the rather banal sentimentality of Aelian's moralizing is generally avoided by Pliny in favour of a more ironic humour. In *HN* 10.52, the friendship between a goose and a philosopher elicits the remark that 'our wiser minds' know the goose for the goodness of its liver.[25]

[22] A remark on the attribution of bad qualities to certain animals, such as greed to the pig, supports this view: 'Men attributed to animals the natural impulses they most feared in themselves . . .' (Thomas (n. 14), 40–1).

[23] In his preface, Aelian calls his work a scholarly one deserving serious attention. However, he admits it is largely unoriginal and is anxious to stress its literary status. He claims elegance of style and a lack of technical vocabulary: contrast Pliny's apology for rustic and barbarian words (pref. 13). In the Epilogue, he claims to avoid monotony by his piecemeal arrangement; he intends to create a 'meadow' or 'garland' with many colours. Such language might put him in the same category as the artistic writers whom Pliny criticizes for their fancy titles and lack of solid content (pref. prol. 24–6). He has no serious philosophical argument on the status of animals to offer. Despite dwelling on their virtues, he appears to take it for granted that they are inferior to man in that they lack *logos* (*De Nat. An.*, prol.: beasts, unlike men, lack reason and speech).

[24] On Pliny's diversity in grammar and style see D. J. Campbell (ed.), *Historia Naturalis Liber Secundus* (Aberdeen, 1936), 5; A. Önnerfors, *Pliniana: In Plinii Maioris Naturalem Historiam Studia, Grammatica Semantica, Critica* (Uppsala, 1956). Similarly, the rhetorical passages, anecdotes, personal opinions, and deviations from the main theme all perform a similar role in eliminating the otherwise dry effect of the factual material (Campbell 33 ff.).

[25] Ironic humour: the following anecdote centres upon the idea of military action being taken in Germany to secure goose-feathers to sell for the making of

On a more serious level, Pliny is sometimes at pains to admire the accomplishments of certain animals at greater length, often suggesting that they possess human qualities. Among those which most arouse his interest are elephants, dolphins, nightingales, and bees. Closer examination of the wording in these passages, however, reveals that Pliny is not claiming that the animal characteristics in question do anything more than resemble certain qualities of man. His admiration of animal accomplishments is remarkable; yet it need not lead to the conclusion that he seriously challenges man's position of supremacy in Nature.

Pliny starts his discussion of animals with the elephant, the largest of beasts on land (8. 1–35). These animals are assigned the 'human' qualities of understanding (*intellectus*, 8. 1) language, of appreciating affection and honour, of possessing such virtues as justice and reverence, and of understanding the obligations of religion. However, in all cases they are said to be 'very close' to human feelings (8. 1) not 'equal' to them. There is a similar vagueness in Cicero's remarks about the distressed elephants at Pompey's games (see sect. 3. B); he mentions a general feeling that the monsters had 'a kind of fellowship' with the human race (*Ad Fam.* 7. 1. 3). It may imply Seneca's distinction (*De Ira* 1. 3. 6) that animals cannot have man's emotions but have certain impulses similar to them. We saw that Philo quoted understanding of God, of national customs, of the state, and the like as peculiar to man's *ratio* (§ 85 Terian), but he also recognized that animals can appear to imitate the attributes of the rational mind (§ 95 Terian; see above, sect. 2. A). This idea of imitation rather than identity is reflected time and again in Pliny's language. He is impressed that elephants respect an 'alien' religion, that of man, in which oaths are sacred, but a qualification is present: 'they are believed' to do this. They throw up grass to heaven 'as though' deputing earth to support their prayers. They possess 'a kind' of insight into justice. They also possess memory (8. 15), a quality found in some animals according to Aristotle, but not implying *ratio* as well.[26]

pillows—effeminate luxury according to Pliny. 'For this reason, the commanding officers of auxiliary posts frequently get into trouble for sending whole cohorts away from sentry-outpost duty to capture these birds' (10. 54).

[26] The passage on dolphins (9. 20 ff.) should be mentioned. It contains many fascinating stories but displays the same qualifications in essential phrases as did

The story of the elephant which could write a Greek sentence (8. 6) has been quoted as evidence[27] that Pliny thought animals were equal in intelligence to man, since Aristotle said that writing was an attribute only of man. Pliny clearly sees it as a mark of intelligence and teachability. However, while the animal apparently wrote frequently, and not as an isolated instance, there is nothing to suggest that it could compose any other sentence. This feat would therefore fit Seneca's definition of the mechanical nature of animals' behaviour (above, sect. 2. A). There is no variety in their so-called *ars* (*Ep.* 121. 23). When commenting on the beauty of the nightingale's song, however, Pliny seems deliberately to assert that this is *ars* in the human sense: '. . . lest anyone should doubt that this is a matter of art, each bird has several songs and not all have the same but each has songs of its own' (10. 82). Yet it becomes clear that there is something too mechanical in this *ars*: '. . . yet these exceptional and artistic trills gradually cease after a fortnight, but not in such a way as to suggest the birds are tired or have had enough. Subsequently, when the heat has increased, their song changes completely with no modulations or variations' (10. 85). Such limitation would not be a feature of human *ars* inspired by rationality. Moreover, the mention of a change in temperature suggests that the song was stopped by the arbitrary outside intervention of *Natura*. This being so, *Natura* could also be the author even of such a varied song. Indeed, there is no reason why her *varietas* should not sometimes be apparent even within one species.

There are similar limitations to the learning ability of talking magpies. They need to hear a word frequently or else they forget it (10. 82).[28] Pliny is also interested in the fact that one nightingale may teach another to sing, a story found in Aristotle (*HA* 4. 9, 536^{b}17). Certainly, this suggests an element of memory (above, sect. 2. A) in the bird. Once again, though, Pliny does not

the other examples, e.g. when one of their number was captured, other dolphins exhibited 'a sort of grief which could be interpreted as suing for compassion' (9. 33).

[27] M. Vegetti, 'Zoologia e antropologia in Plinio', in *Plinio il Vecchio*, 17–31.

[28] Even the most apt species learn only 'in the first two years of their life'. Popular belief, according to Pliny, limits the species capable of learning to those 'who have five digits on their feet'. This latter is perhaps due to a feeling that in this respect they resembled man.

allow the 'human' comparison to be too exact; he talks of 'a sort of criticism' on the part of the teacher (10. 83).

An underlying idea in this passage is the miraculous artistry of *Natura artifex*. Indeed, the craftsmanship of Nature is the prominent theme of a number of these passages in praise of animals. Just as she managed to put so much into so little in the case of insects (11. 4), so too the small size of the nightingale (10. 81) is amazing when compared with her singing capacity: '. . . so great a voice in so tiny a body'. Nature has placed 'in so small a throat (*parvulis faucibus*) all the devices which human *ars* has contrived with the ingenious mechanisms of the *aulos*' (10. 82). It is interesting that we are told of insects that they possess 'outstanding gifts of nature; cleverness, mind, art' (11. 7). It is not so much the degree of these qualities as the fact that they are present at all in so small a creature which interests Pliny. Here, once again, he shows his preference for recording a miracle of *Natura* over investigating how these qualities really work and hence how they really compare with man's. It is his 'purpose to point out the manifest properties of objects, not to search for doubtful causes' (11. 8).

Of all the insects described by Pliny, bees arouse the most admiration. There was a general idea that these creatures possessed some mysterious superiority.[29] Pliny describes their organization and talks rhetorically of their *ratio*. 'Which men, I declare, could we compare in *ratio* with these creatures, who are, in this respect at least, superior, that they recognize only the common interest' (11. 12). We may note, however, the qualification 'at least' and the fact that the 'recognition of a common interest' may mean no more than a particular kind of survival instinct. As with insects generally, everything is being brought into relation to their small physical size. This is made clear when Pliny says:

[29] Virgil's language in *G.* 4. 219 is essentially a poetic image of the greatness of the divinity pervading the universe and all its creatures. It cannot be pressed on points of exact philosophical doctrine. It is certainly not clear why he should then single out bees alone from the rest of creation as being 'partem divinae mentis' (220–1). Perhaps there is a hint of that preoccupation of Pliny's with insects generally and with bees in particular: so many gifts encased in so small a body. The proportion of 'divine spark' to body would be larger, cf. *Aen.* 6. 724: the union of mind and universe produced all creatures but the divine seed is deadened by their sinful bodies. This, together with their organization, etc., makes them remarkable. Cf. also the play on *tenuis* at the beginning of *G.* 4, and 'levium spectacula rerum' in l. 3.

'after that [referring to their tiny amount of blood], let us estimate their *ingenium*' (11. 12). Rather than suggesting that the bee is superior to man, the overall tone of Pliny's passage suggests that much of his interest is centred on its usefulness to man. He begins by saying that bees are rightly the most admired of insects because they alone of that genus have been created for the sake of man (11. 11). They collect honey, the most healthful of juices for man (*saluberrima*), and they make wax, 'that has a thousand different uses'. The 'human' qualities they possess complete this passage. The overall impression is not that Nature has created an animal equal to man: rather, the apparent similarities make a more harmonious bond between man and his efficient servants.

In all these passages, enthusiasm tends to mask the qualifications in Pliny's language. There is no evidence that he really identified the qualities observed in animals with the human intellect and the human emotions. To end this section, something should be said about the powers of prognostication thought to be possessed by some animals. Here, above all, it would be easy to claim some superior intellectual gift for animals to explain why they anticipate events which man cannot foretell.

In a general statement on the subject, Pliny starts by saying that 'Nature has bestowed on many animals the faculty . . . of prognostications' (8. 102). This emphasis on the power as a gift of *Natura* is reminiscent of *HN* 8. 10 (instinctive fear of predators) and 27. 7 (medical knowledge of animals). In the latter, as we saw, this direct link between Nature and animals is a compensation for the absence of *ratio* rather than an indication of the presence of something better. Certainly, there is no question of animals' possessing an intellectual power greater than man's. Any superiority lies in their physical senses. Seneca said that animals' senses are better than man's although they lack *ratio* (*Ep*. 124. 4). The power of *praesagia* given to certain animals need not be seen as a mysterious prophetic power. As Virgil said of crows' ability to sense changes in the weather, 'it is not, I think, because they have a greater talent by divine gift or greater foreknowledge by fate' (*G*. 1. 415–16). They are just extremely sensitive to change in the atmosphere (1. 417–22). That Pliny attributes *praesagia* to nothing more than superiority in the physical senses is suggested by the example he quotes of the fox.

The *sollertia* of the fox consists in its putting its ear to the ice to test its thickness (8. 103). It is superior to man only in the acuteness of its hearing.

As we saw earlier (sect. 2. A), Seneca stressed the self-preservatory nature of animal instinct, and indeed the implication in Pliny's story is that the fox was making sure that the ice was thick enough to bear him by listening for water. In *HN* 18. 208, such motivation is made explicit in the case of animals' weather prognostications. Like man, animals, 'despite their greater *sagacitas*, since their lives depend on it', can be deceived by unexpected weather disrupting the seasonal pattern outlined by the stationary stars. Such animal powers are, then, a logical necessity and lack any overtones of mystery. Indeed, Pliny's point in *HN* 18. 208 is that they are not infallible. Even animals, with their greater *sagacitas* in such matters, can be taken just as unawares as man by unpredictable weather changes that are not dictated by fixed stars. (Those dictated by the fixed stars, it will be recalled, worked on a rational principle and hence were observable and predictable by man; see Ch. I. 4.)

For Pliny, such prognostication forms an alliance between animal and man—*societas* (8. 102)—but the similarity is generalized and not exact. Man's *ratio* and *observatio* can foretell natural events which follow a regular pattern; animals have a keener awareness of the changes in Nature by means of their superior physical senses alone. Yet even they can be taken unawares. Such physical superiority is a compensation for lack of *ratio* and is an aid to self-preservation. It enables them to escape disasters which threaten to destroy their environment. The mice who escape through advance migration the imminent collapse of buildings (8. 102) perhaps discern tremors too slight to be noticed by man.

Societas, affinity between animals and man, exercises an undeniable fascination on Pliny and on Romans generally. Such affinity might excite enthusiasm for animals' skill, as in many of the examples from the *HN*, or sympathy for their 'emotional' reactions to their plight as described in Cicero's letter on Pompey's elephants (above, Sect. 1. A, cf. 3. B). However, we are not justified in attributing to this interest any greater significance than that of analogy and resemblance. There is no intellectual equation to be made. Where animals are superior, it is in natural instinct, physical senses only. For the Stoic thinker in particular, there remains an unbridgeable gap between man's *ratio* and the impulses of animals.

3. *The Cosmic Order: Man, Beast, and* Natura

Finally, let us consider the significance of animals in their role as a part of *Natura* and as instruments and manifestations of her power. At the end of sect. 2. A, it was suggested that the relationship with *Natura* of creatures which do not possess *ratio* is more straightforward than that of the rational animal man. We move from a comparison of man and animal as fellow creatures to an appreciation of animals' wider cosmic significance and its relation to man's life in Nature.

A. TRADITIONAL LINKS: ANIMAL PORTENTS AND COSMIC SYMPATHY

In his discussion of the prognostication of weather and danger by animals, Pliny briefly mentions the function of animals as portents for man of impending upheavals in his social structure. Extispicy is the particular form mentioned here and Pliny pauses only to mention the fascination it exerts on a sizeable proportion of mankind before returning to the environmental prognostications described in sect. 2. A. The difference is that extispicy is devised by man, the animal playing a wholly passive role whether it is alive or dead. It is not, therefore, connected with the animal's individual endowments, physical or intellectual. Instead it is connected with cosmic sympathy, with the idea of macrocosm and microcosm. The animals here act merely as the instruments of *Natura*, reflecting on a smaller scale the upheavals portended by the larger cosmic phenomena like comets and earthquakes. The principles behind these phenomena are obscure (2. 97) but they can be viewed generally as evidence of the coherence and sympathy of the whole living cosmos.[30] As smaller manifestations, the results of extispicy are not, perhaps, to be taken too

[30] The living cosmos: *SVF* ii. 633 (D.L. 8. 142–3), the world is a living being, rational, animate, intelligent; *SVF* ii. 635 (Philo), the whole heaven and earth is a living, rational being; *SVF* ii. 638 (Galen), the cosmos is a living being . . .

seriously. Pliny's brevity on the subject of extispicy's devotees takes on an ironic note when viewed in the light of criticism from some Stoics of such divination on the grounds of triviality[31] and Pliny's own reservations on physiognomy as discussed in Ch. III. 2. B.

After the remarks on extispicy, Pliny mentions, again briefly, another form of prognostication by men from animals, augury. He is prompted by the mention of the mice who anticipate the destruction of their surroundings to add that 'augury has been established as a science at Rome, with a priestly college of the greatest dignity' (8. 103). Like the remark on extispicy, this short comment too may be mildly ironic; do small rodents (*musculi*) really warrant an august priesthood? The context of the remarks suggests that the *sollertia* of the mice and of the fox in the following example makes them particularly apt animals for augural observation. They show greater sensitivity to atmospheric and other changes in their local environment. Hence they may also be more closely attuned to those cosmic disturbances which are reflected passively by animals in extispicy. Further on, in *HN* 10. 34–5, Pliny remarks on the fact that the appearance of the screech-owl is regarded as a bad omen, especially at public auspices, but then adds cases where its omen of doom has not been fulfilled (see below). It may be that, while Pliny believes in the mysterious powers of cosmic sympathy, he refuses to attribute miraculous supernatural powers to the animals themselves. He remains aware of the capacity for error of animal *sollertia*, even with regard to weather. There is all the more reason not to trust them blindly when they are held to sense a greater and more mysterious turbulence in nature.

Although he agrees in principle with the idea of cosmic sympathy,[32] Pliny avoids categorizing *divina Natura* too minutely (Ch. I. 4). His universe is more than the sum of its parts. The

[31] Cic. *De Divinatione* 1. 118: see Ch. III. 2. B and ibid. n. 53, cf. also Pliny's non-committal 'it is said' that the filaments in the tiny livers of mice correspond to the number of days of the moon in the month (11. 196). He gives without comment details of various deformities of the liver found by extispicy (11. 189 ff.). Cosmic sympathy at least makes such a correspondence a possibility. Not so the discovery of snakes in a victim's entrails (8. 197): this, says Pliny, would seem credible only 'to those who believe that, on the day Pyrrhus died, the heads of his victims when cut off crawled around licking up their own blood'!

[32] e.g. 24. 1–2. See also Ch. VI. 2. C.

links between man, beast, and *Natura* are more complex than such mechanical parallels suggest. Pliny recounts many other strange occurrences in the animal world that were generally regarded as ominous or portentous. In his opinion, however, they need not necessarily have any wider implications with regard to man.[33] Animal oddities may be simply *mirabilia naturae*. Pliny generally avoids superstitious interpretations of animal deformities, no doubt because *varietas naturae* could account for them.[34] He shows little interest in the beliefs about the animals Aristotle called 'dualizers',[35] such as the bat, a mammal which flies like a bird. He is, however, aware that uncharacteristic behaviour in an animal is a cause of superstitious alarm. This is particularly so with a creature like the screech-owl whose usual habits are themselves strange. It is a night bird whose haunts Pliny describes as being 'not merely unfrequented but actually terrifying and inaccessible' and its cry as 'not a song but a moan'. Consequently, when it appears in inhabited places or in daylight, it is *dirum ostentum*, an unnatural intrusion of the unfamiliar and wild into civilization. Yet even this need not be significant. Pliny says he knows of many cases (*plurium*) where its appearance on a roof has not been fatal to the building's inhabitants (10. 34–5).

It is true that the records of animal portents in the *HN* are too numerous to list. Yet on the whole it seems that, for Pliny himself, the occurrences whose unnaturalness or perversion holds some significance for man are those which involve man's misuse of *Natura*, not the oddities which appear in Nature spontaneously. This may be seen in *HN* 8. 55, where Antony's yoking

[33] For example, if the snake found to have swallowed a child (8. 37) was popularly regarded as a portent, Pliny does not bother to mention it. He records it solely as a marvellous occurrence.

[34] Arms of different lengths and extra fingers in human beings are mentioned without comment (11. 244, 245). But extra limbs are perhaps unnatural because they are unnecessary. An Egyptian with an extra pair of useless eyes at the back of his head is called a *portentum* by Pliny (11. 272), but without any apparent religious significance attached.

[35] See Lloyd (n. 9), 45 ff. While Aristotle gave rational explanations of such creatures, the traces of earlier feelings about them can be seen in his occasional references to them as anomalies or natural deformities. There is possibly an echo of a similar idea in Pliny (11. 255): pigs are thought to belong both to the class of solid-hooved and the class of cloven-hooved animals. Consequently, their pastern bones (according to Pliny normally associated with cloven-hooved animals; the rhinoceros is an exception) are misshapen (*foedi*).

of lions to a chariot is classed as a *prodigium* which, Pliny specifically points out, reflected the subjugation of noble-hearted men in the Civil wars (see below, 3. A). Profitable examination of Pliny's cosmic order and the significance of animals in it rests above all upon this question: what constitutes use or abuse of animals by man?

B. CRUELTY: A FACET OF *NATURA*

The complexity of the Romans' attitude was mentioned in the introduction to this chapter. They could keep pets at the same time as watching the slaughter of beasts in the arena with equanimity and enjoyment. In general, a concept of cruelty or kindness with regard to animals did not develop in classical Rome. Limits are set but 'not where we would set them'.[36] The notion of kindness to animals was a comparatively late development, even in more modern times.[37] The only appreciation of 'one of the most outstanding paradoxes of the Roman mind'[38] is to be found in a Greek, Plutarch. 'Kindness', he says (*Cato Maior* 5), 'has a far wider sphere than justice, which is confined to men. Kindly men will take good care of old dogs and horses.' In his *De Sollertia Animalium*, he concludes that there should be no wanton cruelty or luxury in man's dealings with animals (965 A–B). It is noteworthy, however, that the centrality of man is never disputed: a little earlier in the same dialogue (959 F), one speaker condemns the games because they encourage man's insensibility, but his interpretation of the Pythagoreans' practice of kindness to animals shows that he is concerned with the insensibility of man towards man: he claims that they sought thereby to inculcate philanthropy. A similar point is made by Plutarch himself in *Cato Maior* 5: we ought to cultivate kindness and consideration towards animals if only for the sake of gaining practice in humanity, for when we come to deal with our fellow men.

For the Romans, however, the games served an important political function and were an expression of Roman power.[39]

[36] K. Hopkins, *Death and Renewal* (Cambridge, 1983), 28.
[37] Thomas (n. 14), esp. 150 ff.
[38] Toynbee (n. 2), 21.
[39] Cic. *Pro Sestio* 106: the arena is one of the places where the Roman People's views are most clearly expressed. In the Empire, with the emperor the centre of

In particular, there was a strong connection with the military ethos. There were special connections between the army and the slaughter of animals in the arena. Hunting was recognized as a useful training for war. Men of the praetorian guard gave exhibitions of beast-fighting under Claudius and Nero. In the third century, the praetorians had their own hunters and game-enclosure.[40] Even outside the official arena, the fortuitous appearance of a large and formidable-looking animal could not be ignored. In *HN* 9. 14–15, Pliny reports that the appearance of an *orca* (the identification as 'killer whale' seems unlikely) in the harbour at Ostia caused Claudius to organize an impromptu *spectaculum* with the praetorians attacking the beast: 'An *orca* was actually seen in the harbour of Ostia being assailed by the emperor Claudius . . . He ordered a barrier of nets to be stretched across the harbour and setting out himself with the praetorian cohorts, he gave a *spectaculum*, the soldiers throwing lances from the warships . . .'. Once the larger and fiercer African beasts were regularly imported for shows and games, they seem almost entirely to have superseded the older *venationes* of Italian animals which might have had religious connections.[41] *Venationes* 'with-

attention and in control of the combatants, the whole games process 'was an enactment of imperial power' (Hopkins (n. 36) 18). Augustus saw fit to record that, in 26 *venationes* of African beasts, about 3,500 animals were killed (*Res Gestae* 39–42).

[40] Military ethos: Jennison (n. 2), 141; Suet. *Claud*. 21; Dio 61. 9. 1; *ILS* 2091. In 363, Julian's cavalry came upon the *vivarium* of the king of Babylon. Their reaction was to break down the gates and kill the lions, bears, and boars with javelins and other missiles (Amm. Marc. 24. 5. 2). It should also be recalled that the gladiatorial contests between man and man were cited as providing an edifying display of various military virtues, e.g. Pliny, *Pan*. 33. For Cicero, gladiatorial combat was like military training in that it offered a lesson in endurance and self-discipline (*Tusc*. 2. 41). While he had sympathy with those who thought such contests 'cruel and inhumane' (41), he had no qualms about the use of condemned criminals as gladiators.

[41] On the older *venationes* of Italian animals see Jennison (n. 2), 42. For the importation of African beasts, see 8. 64. This account of an attempt by the senate to stop such importation in fact bears witness to the growth of the practice from an early date. Pliny tells us that the resolution was 'ancient' but that it was partially repealed by a resolution carried by one Cn. Aufidius, tribune of the plebs, in the popular assembly. The dates remain uncertain, but, on the whole, the trend towards such imports seems clear and is confirmed by the rest of the passage: the number of 'African' beasts displayed gets steadily larger. Scaurus exhibits 156 leopards in 58 BC, Pompey 410, and Augustus 420 at once. The scope for such importations did of course increase as the Empire expanded from the 2nd c. BC onwards.

out African beasts' were taking place in Varro's day (see below) but Ovid (*Fasti* 5. 371) draws attention to the use of 'unwarlike' animals of local origin such as hares at the *venatio* in honour of Flora, which suggests that the practice was by then unusual.

It is true that a well-staged hunt, which concentrated on skill rather than on the sensationalism produced by the exhibition of fierce and exotic beasts, could evoke a certain aesthetic pleasure in the art displayed. Varro had commented on the sight of an 'Orpheus' calling beasts in an enclosure for the guests' amusement: 'It seemed to me to be no less attractive (*formosum*) a sight than when the hunts of the aediles take place in the Circus Maximus without African beasts' (*RR* 3. 13. 2–3). However, the more savage beasts appealed to a widespread desire for the combative aspect of such spectacles. One of the spectators in Plutarch's *De Sollertia Animalium*, Soclarus, justified beast-fighting as harmlessly channelling man's aggressive instincts by the display of skill and courage directed against witless force and violence (959 c).

Several of Pliny's stories reflect this appreciation of combat central to Roman enjoyment of the games. In one passage, it is the elephant which delights Pliny by its skill. The wounded beast 'in a marvellous conflict' crawled on its knees and threw its human opponents into the air 'as though they were being thrown by artistic skill, not the fury of a wild beast' (8. 20). The crowd was delighted. He then tells of a particularly skilful killing of an elephant by a javelin in a single blow under the eye. What Pliny and the crowd appreciate are conspicuous examples of the qualities of a good fighter. In Pliny's story of Alexander and the fighting hound, the language used to describe both the combatant dog and the royal onlooker, an experienced military leader, is significant. When the dog refused to fight an opponent it disdained, Alexander's *generosus spiritus* was offended and he killed it. He was sent another, with instructions to pit it against a lion and then an elephant. 'No show gave him more delight', says Pliny, and the *artifex dimicatio* in the attack on the elephant is stressed (8. 149).

The phrase *generosus spiritus* is precisely the same as that later used to describe the lion in *HN* 8. 48: '. . . the most noble-spirited of all animals, lord of all the rest . . .'. In *HN* 8. 50 ff., Pliny describes its *generositas* and *nobilitas* (see also below, 8. 55, 3. c).

The character-distinction made by Aristotle between animal species (*HA* 8.1, 588^a18–25) had become popularized. We saw that Pliny is often careful not to say that such characteristics are identical with man's (2. B). Seneca had criticized the popular tendency to identify such characteristics with those of men because it implied that they possessed the same moral and rational qualities as man (see 2. A). In *De Ira* 2.16.2, he criticizes in particular the popular esteem assigned to fierce animals which are praised for their nobility: is not patience, the quality of the ox, more admirable? His main purpose is to oppose the tendency to take the analogy with men's characteristics to the point where fierce races of men are praised above all others; man should live by *ratio*, not by *impetus*. For the present purpose, however, the passage confirms the popular estimation of certain animals as *generosus*, especially lions and elephants. These appear in *De Clem.* 1.5.5 as stock examples of *ferae generosae*. Cicero's expresson of pity is directed at elephants. Statius commemorates the death of a brave lion in the arena, comparing it to a gladiator (*Silvae* 2.5).

Cicero's mention of the effect made by the elephants at Pompey's games has interesting parallels in both Pliny and Seneca.[42] These animals, as becomes clear from Seneca's account (*De Brev. Vit.* 13.6), had been pitted against human criminals. In Pliny's account, the apparent supplication of the elephants for pity is the point of interest. It so affects the crowd that it curses Pompey, 'forgetting the general and his munificence so carefully devised for its honour' (8.21). Pliny adds that Pompey eventually 'paid the penalty'. The asking for and granting of *clementia* is as much a part of the fighting ethos as is the skill and bravery of combat. The lion is exceptional among animals for showing *clementia* to suppliants (8.48). Pliny's remarks suggest that the human general, Pompey, who, as organizer of the games was indirectly responsible for the elephants' attackers, should have shown *clementia* to the suppliants in this fight.

Pliny's interest is above all in the skills and qualities of a good contest. Despite his humanitarianism, he shows no interest at all

[42] The story is also found in Dio 39.38.2–4, who agrees with Pliny that the elephants were killed. Seneca seems to be alone in thinking that they trampled the men to death.

in the criminals pitted against the animals. In this, he differs from Seneca, whose emphatically anthropocentric Stoicism allows him no interest in the elephants. His condemnation of Pompey is solely on the grounds of his inhuman treatment of the men pitted against the beasts.

It can be seen from these examples that the idea of sympathy for these animals is ambiguous. Even in Pliny's case, interest is aroused primarily by the display of fighting skills and qualities rather than by the plight of the animals *per se*. For Seneca, man is unambiguously the only combatant of importance. A later Stoic writer, Marcus Aurelius, appears from his personal writings to have had no feelings at all about such contests, other than that of boredom (*Med.* 6.46).[43] The plight of the animals certainly does not affect him. He mentions the mangling of the human beast-fighters, but even this is factual rather than sympathetic (*Med.* 10.8.2).

Other attitudes frankly show a taste for the shocking or the bloodthirsty rather than for the skilful elements involved. The sensational tone of Martial's *De Spectaculis*, the gory detail in *Ep.* 7, and the subject of *Ep.* 12–14 (the killing of a pregnant sow) go beyond the needs of rhetorical vividness by modern standards.

Notions of cruelty and feelings of compassion towards animals were not, then, very developed concepts in classical Rome. Animal games catered for a love of *mirabilia* in the form of the size, number, exotic provenance, and fighting qualities of the beasts. Pliny, as a government official, acquainted with the political aspect of such games, would be unlikely to feel sentimental. Indeed, in the *HN* cruelty is a facet of *Natura* and one often present in her *mirabilia*. Her darker, savage side was discussed in Ch. I.4 and will be explored further in Ch. V. In the *HN*, Nature at play, *Natura ludens*, is not a cosy image. We may go further: man, in entertaining himself by means of contests

[43] This is not to dispute his aversion to cruelty to men but there does not seem to be sufficient evidence to suggest that it extended to animals. In Dio 72.29, he gives gladiators blunt harmless swords against each other and opposes performances by a *man-eating* lion. The passage in the *Historia Augusta* 4.4.8 says that he would sometimes let himself be taken to the theatre, hunts, or *spectacula* as a concession; his reluctance could just as easily be due to boredom as to any objection to the cruelty involved. See now R. B. Rutherford, *The Meditations of Marcus Aurelius: A Study* (Oxford, 1989), 117.

between beasts, does no more than imitate *Natura*. Certainly, there is a continuous war of survival in Nature. Natural enmities between animals of different species might be explained by the need for *Natura* to keep a balance. In *HN* 8. 91, for instance, *Natura* gives the crocodile more than one enemy because it is so great a pest. All animals, according to Pliny, know not only what is advantageous for themselves, but also what is detrimental to their enemies. Yet Pliny sometimes feels no need to offer even this justification. In *HN* 8. 34, the enmity between the snake and the elephant is arranged by *Natura* for no other reason than 'to provide a *spectaculum* for herself with a pair of contestants'. Nature provides for herself the entertainment of the arena. The slaughter of one animal by another is related by Pliny in a detached and factual manner. Killer whales attack and kill another species of whales, especially the pregnant females and young as they are the more vulnerable (9. 13). Yet this too is depicted as a contest watched by spectators: *spectantur ea proelia*. The size and ferocity of the killer whale are particularly fascinating to Pliny: '. . . their appearance can be represented by no other description than that of an enormous mass of flesh with savage teeth' (9. 12).[44]

Man's interest in the great size and ferocity of certain beasts is seen by Pliny to be no more than an imitation of *Natura*. *Natura* herself is a theatre full of *mirabilia*. Animal fights are only one element in Pliny's imagery of *Natura* as providing a *spectaculum*. Some of these concern activities of men involving animals; they may provide more evidence as to the parts played by man and animal in the setting of *Natura*.

[44] The *Cynegetica* of 'Oppian' (probably of the 3rd c.) is a particularly striking example of literature reflecting a fascination with the skill, bravery, and sheer violence of the contests between man and animal and between animal and animal. God has given the beasts 'strength, courage, and wits' not far inferior to the human hunters' (4. 8–9). The hunting narrative is interspersed with scenes of gory combat between animal and animal. The battle of the deer and the snake makes particularly vivid use of analogy with the games. The deer swaggers and almost smiles (2. 237, 246; cf. 4. 189–90, 200, where lions are compared to a surrounded warrior and a prize boxer). Bloodthirsty sensationalism recurs even in the passages on the human-like love to be found among animals: the jealousy of the male wild ass towards its young is enlarged upon with the aid of 'human' parallels from the most lurid myths, culminating in that of the Thyestian feast (3. 198–250).

C. PLINY'S *SPECTACULUM* IMAGERY: VICE AND MISTREATMENT OF ANIMALS

The idea of *Natura* as being a theatre and providing a spectacle fits in well with the Stoic idea of man as a contemplator of Nature, discussed in Ch. I. 4. It is a particular feature of the popularized Stoicism of Epictetus, who takes up the older idea of life as a fair[45] to express his view of Stoic contemplation. One must look, not superficially, but with a view to understanding the wider scale of things (*Discourses* 2. 14. 23–9). Life is God's pageant and holiday which man is invited to view and join in (4. 1. 104). A somewhat different but related idea is that found in the *Meditations* of Marcus Aurelius. Past history is a staged drama which repeats itself with different actors (*Med.* 10. 27). This may be related to his remarks on the monotony of games: 'they are an everlasting repetition of the same sight: such must be the effect of the whole of life' (*Med.* 6. 46). Life is a rather monotonous series of *spectacula*.

Pliny's *spectaculum* imagery may also be seen against the background of the general Roman interest in the 'set scene', as for instance in their theatrical ideas of views and prospects both horticultural and natural. Pliny himself saw theatrical design in vistas of mountains and cities (4. 30).[46] Conversely, art imitated nature in the 'natural' setting contrived in the arena for the hunt spectacles.[47]

It has been observed[48] that Pliny saw much in the cruel and bizarre facets of Nature that mirrored the sights of the amphitheatre and the stage. Theatrical imagery is used of the aggression between animals and of the vices of men. Besides the battle of the whales already mentioned, there is the fight between the osprey and the fish (10. 9, 'a duel worth watching'). Fishermen summon the help of dolphins to catch mullet 'for the denouement of the show' (9. 30). Theatrical images, when translated to the human sphere, are usually connected with man's pleasure,

[45] Cic. *Tusc.* 5. 9 attributes it to Pythagoras, cf. D. L. 8. 8.

[46] See Ch. II. 5 on gardens.

[47] See P. Grimal, *Les Jardins romains à la fin de la République et au deux premiers siècles de l'empire*[2] (Paris, 1969), 335–49 and in particular 341 n. 4: also R. Auguet, *Cruelty and Civilisation: The Roman Games* (London, 1972), 99 ff.

[48] See especially M. Vegetti (n. 27), 117–31.

cruelty, and vice. Displays of greed, for example, may have a theatrical element: the colour-changes of a mullet dying at table may be watched if it is placed in a glass bowl (9. 66). The 'theatrical' behaviour of the powerful may have bizarre aspects: Pompey's *mirabilia* in his theatre follow a comment on hermaphrodites (7. 34). Antony uses lions in a chariot (8. 55). Extremes of behaviour may be attributed to professional actors: Clodius Aesopus ate song-birds (10. 141–2).

Two points of interest arise out of these observations. Does Pliny mean to suggest that there is a straight parallel between the aggression of animal towards animal and the vices of man? Do those passages which involve the use of animals by man suggest that man's vice includes the cruelty to the animal as such? The answers have to some extent been determined by the conclusions reached earlier in this and the preceding chapters. If we believe that Pliny considers man to be unique in his possession of *ratio*, any comparisons between man and beast are going to be more complex. It is by misuse of *ratio* that man indulges in vice, so that vice is logically denied to animals. Their cruelty to each other is an amoral fact of Nature in the *HN* (sect. 3. B). The answer to both parts of the question lies in regarding animals not as directly parallel to man but as the passive instruments of *Natura* as suggested earlier (2. A, *fin*.). Man's vice is a result of his use of *ratio* to pervert *Natura*. We must emphasize the bizarre, perverted quality of those *spectaculum* images involving man using animals. We shall see that his behaviour is a perversion of the facet of nature which the animal embodies. It is not to Pliny an instance of cruelty to animals *per se*, and is not indicted as such.

Good examples are the passages on Mark Antony and Hanno the Carthaginian and their use of lions. The vice here is lust for power. Both attempt to display their own power by perverting rather than just imitating *Natura's*. Lions were the chief of those animals popularly considered to be 'noble' (sect. 3. B). They might also be said to embody the untamed power, the majesty, and the ferocity of *Natura*. Indeed, Pliny at one point refers to the lion as ferocity personified, *feritas* (8. 55). The arrogant and perverted nature of Antony's lust for power in yoking lions is signified by Pliny's comment that it was a *prodigium* to place 'generosos spiritus' under a yoke. When he rode this chariot in the ignoble company of the actress-courtesan Cythera it was a prodigy which eclipsed the natural portents of a disastrous period (8. 55). The Carthaginian leader Hanno was the first man to tame a lion and this, says Pliny, contributed to his political downfall. His action, which amounted to a paradoxical claim to have tamed *feritas*, was culpable audacity (*ausus*, 8. 55). It may be compared with the interference with Nature's secrets by the inquisitive in *HN* 2. 87, where Pliny says that *ratio* offers an opportunity for impudence. Hanno, too, made over-ingenious use of *ratio*: the Carthaginians felt their liberty threatened by a man of such ingenious mind (*tam artificis ingenii*).

The *spectaculum* image provided by the mullet dying at table (9. 66) is an example of man's perverted ingenuity in luxury rather than of cruelty to the fish *per se*. Pliny goes on to say that the gourmet Apicius, 'a man with a natural aptitude for every kind of luxurious device', devised a method of killing and cooking fish in a sauce made of their own companions (*in sociorum garo*). This anthropomorphic touch is not to arouse sympathy for the mullet but horror at the man who devised such a death. Only man could think of killing another animal by means of its own species since he alone was accustomed to this type of slaughter. 'Even sea-monsters and fish are cruel only to other species. But in man's case, on the contrary, most of his evils come from his fellow men.' The mullet is the instrument of *Natura* on which man exercises his perverted ingenuity.

The quotation of *HN* 7. 5 suggests a further significance. Man's maltreatment of animals symbolizes his maltreatment of his own species. Pliny's humanitarian philosophy would suggest that man's hostility to or misuse of man is the most unacceptable of his perversions of *Natura*. Man can be cruel only to another human being. The description of the actor Aesopus (10. 141–2) who ate songbirds and mimics serves to confirm this interpretation. To eat 'birds who talked with human speech' is described as a surrogate cannibalism: a desire to eat an imitation of a man (*imitationem hominis mandaret*).

The apparently human qualities of beasts are used to give greater rhetorical point to the cruelty of man to man. Pliny is also suggesting that, while it may be a lesser evil to misuse the greatest treasures of *Natura* than to misuse fellow men, there is a logical link between the one and the other. Antony's arrogant treatment of lions reflects his attitude to his human opponents.

Hanno's claim to have 'tamed' the embodiment of the untameable in Nature led to his impeachment by the Carthaginians, 'fearful for their own liberty'. The crowd at Pompey's games curse him for his behaviour to the elephants; the general who ignores human-like pleas for *clementia* is like the eater of 'talking' birds or the tamer of the untameable lion. He may progress from this perversion of *Natura* to an even greater one, the maltreatment of man. Ultimately, Pliny reaches the same conclusion as Seneca, who said of Pompey that he believed himself to be above nature (*De Brev. Vit.* 13. 7). We also reach, in a roundabout way, the anthropocentric justification of kindness to animals found in Plutarch (3. A, *init.*).

The theatre of Pliny's *Natura* is full of *mirabilia*. Many are strange and bizarre. They illustrate different facets of *Natura*: her power, majesty, untamed wildness, even her cruelty. Man the microcosm mirrors these qualities but may often distort them. Sometimes, this is a result of misguided ingenuity. Birds intended by *Natura* to be songsters are killed for food or a fish is killed and eaten in an unnecessarily perverse manner. To cage birds is to flout the law of *Natura*, which had assigned them the open sky (10. 141); it is not called cruelty to animals *per se*. Sometimes man tries to control *Natura*, as when he declares his power by yoking the symbol of *Natura*'s power, the lion. As for the contests of aggression, the subject with which we started, man shows too great an inclination to 'progress' from imitation of *Natura*'s *spectacula* of different animal species fighting to unnatural strife between man and man.

D. THE COSMIC SIGNIFICANCE OF MAN AND ANIMALS

Endowed as it is with many wonderful instincts and senses, the animal is still essentially a passive instrument of *Natura*. Man, by virtue of his unique *ratio*, is free: he can choose whether to follow or to go against *Natura*. He is the centre of the universe, and animals are part of the furniture provided for his use.

Animals are thus in a sense closer to *Natura* than man is, but this state is imposed upon them. When Pliny says that only man has borrowed poison, yet we cannot suppose that animals do not know of these things (18. 3), he makes a moral criticism of man. He does not mean that animals are totally superior since the knowledge that they are given is limited to their survival needs. Just as they were given an instinct for finding and using natural remedies, so, by implication, their knowledge of poison would be limited to knowing what to avoid. Theirs is a sphere of amoral necessity. To man alone, who has choice, is the language of moral judgement strictly relevant.

As her instruments, however, animals must sometimes embody the dark side of *Natura*, as well as the bright, beneficial side which, in Pliny's optimistic view of man's supremacy, normally eclipses it. A few are unambiguously evil, *mala*, scourges of man which induce doubts as to Nature's *benignitas*. Commenting on the snake, Pliny asks whether *Natura* has engendered *mala* or *remedia* more frequently. Perhaps *Natura* has not been entirely malignant even with regard to such animals. Both the snake and the crocodile ('evil on four legs', 8. 89) have as enemies, besides their animal attackers, particular races of men whose scent alone can drive them away (8. 93).

Yet animals do more than divide neatly into *bona* and *mala*. They embody above all the complexity of the relation between *Natura* and her uniquely intelligent creation, man. Those animals which should supply him with benefits can sometimes display a reluctance to do so. Deer, otherwise so mild, show unaccountable *malignitas* (8. 112) and *invidia* (8. 115) by hiding their shed horns, which have medicinal value. Such reluctance is understandable as a survival instinct if the animal is hunted for its parts, like the beaver (8. 109). The horn of the deer, however, like the shed skin of the gecko (8. 111), is a dispensible product apparently unwanted by the animal. Again, certain animals, such as the dog (25. 91), have discovered healing plants, but exhibit *malignitas* towards man by eating such plants only when no human being is watching. These animals represent the ambiguity of *Natura*, the mother and stepmother who makes man struggle harder to claim the gifts she provides for him and thus to assert his supremacy over the rest of the animal kingdom.

Four points may be reiterated to summarize Pliny's thoughts on animals. (i) He follows a tradition of thought which gives the universe a teleological purpose. 'Animalitarianism' as such cannot really be found in his writings. (ii) However, he has a positive interest in animals which would make him an obvious source of interest to later 'animalitarians'. This interest distinguishes him from a near-contemporary writer with teleological and humanitarian views, Seneca, and it inspires his method of stressing man's position by giving due attention to the rest of the creation over which he is supreme. (iii) Such an attitude derives from an aspect of his more popular approach to such questions. He reflects the popular interest in *mirabilia*, of which *Natura* is the provider *par excellence* in all their variety. These include (iv) the excitement and skills of the staged fight between animal and animal. Even fights between animals and men do not appear to trouble Pliny's humanitarian feelings, as they do Seneca's. Pliny records only the wonder of the mob at the elephants' pleas for mercy and the emotions aroused by the human quality of the animals' behaviour. Seneca remarks only on the inhumanity of man pitting other men against such beasts. This is a minority view, that of the stricter philosopher, who considers the recording of *mirabilia* without obvious moral purpose to be unworthy of a serious writer (*Brev. Vit.* 13.3, 9; 14.1).

With the assessment of the relation of man, beast, and *Natura*, the picture of man's position in the cosmic scale is completed. The final chapters will consider his activities in *Natura*, the uses to which he puts his environment. Despite his supremacy, the ambiguity of *Natura* the mother and stepmother remains. It is, to a certain degree, reflected in the division of the next chapter: man on the land and man on the sea.

V

Land and Sea

Introduction

Animals in conflict represented for Pliny the idea of a fundamental struggle inherent in *Natura*. *Naturae dimicatio*, the conflict of nature, is also to be seen on a much larger scale in the unending conflict between the opposing elements, land and sea. In this struggle, the sea dominates, 'snatching' and 'devouring' land: 'Ocean . . . in its greedy progress (*avido meatu*), overwhelming all the lands which shrink in terror at its coming, washes also those which resist with a winding and broken coastline' (3. 5). The sea 'bursts in' and 'swallows up' the breadth of the Peloponnesian peninsula at the Isthmus of Corinth (4. 9). Such encroachment may be seen as an instance of a 'peculiar jealousy (*invidia*) on the part of *Natura*, which indulges the sea's greed without any limit' (6. 1). The land resists (*invitis . . . terris*), as is proved by the narrowness of the passages thus created (6. 2). The idea is best summed up in *HN* 2. 170 ff. Here the inequality of the struggle is suggested, since the seas which 'encircle the globe and divide it in two, depriving us of half the world' are vast beyond measure: 'the holding owned by such a mighty mass is bound to be enormous and without limit'. While the sky, by creating uninhabitable zones, has 'stolen' three-quarters of the earth, 'the extent of ocean's pillage is uncertain'. In contrast, Pliny moralizes on the theme of earth's insignificance (174, see 1. A. i).

The sea is 'uncertain': as such, it is threatening, not only in its size, as in *HN* 2. 172 above, but also in its unpredictability. It is an ambiguous element of nature. When Pliny disinguishes between different types of natural causes (Ch. I. 4), he defines as unpredictable the effects of sublunar reactions of cloud or air. Besides steady, predictable *venti*, there are irregular gusts, *flatus* (2. 116). Such happenings belong to a darker, inexplicable side of *Natura*. The description *bruta*, 'senseless', suggests that here *ratio naturae* is at its least discernible. The combination of air and water poses a particular threat to the sailor; the danger posed by sudden squalls in the Mediterranean was remarked upon by many Latin authors.[1]

The sea is 'the most savage area of Nature' (36. 2). In contrast, *benignitas naturae* manifests itself exclusively in *Terra*. In 2. 154, Pliny says that *Terra* alone is never angry with man (*irascatur*). Water makes mists, waves, and torrents, and air makes clouds and storms, but earth is 'kindly, gentle, benign, always a servant to man's needs' (2. 155). Even where the darker side of *Natura* seems to manifest itself on land, there are attempts to preserve the idea of *Terra* as man's friend (see Ch. I. 3). Volcanoes, for example, are described as *Natura*'s savage side shown through the medium of fire, not earth, which is itself threatened by this activity: 'Nature raves savagely, threatening the lands with combustion' (2. 236). Even the difficulties met with in the cultivation of the earth are ingeniously described as proof of the bond between man and *Terra*. For example, Pliny excuses *Terra* for being poor in some places: she makes the faults of the soil obvious so that man will not be uninstructed (17. 32). The struggle with wild Nature which must take place if cultivation of the earth is to be successful is represented by Pliny as a battle in which man is allied with earth against the brambles which try to engulf her: '. . . brambles, which are poised to take over everything, unless resistance is offered by civilization, so that it would be quite possible to believe that man was created for the sake of the earth' (17. 96). According to Pliny, earth is the one division of Nature on which men have bestowed the deserved title of 'mother' and she belongs to them as the sky does to God (2. 154). When he takes to seafaring, however, man is deserting the natural element assigned to him. In themselves, these are unexceptional elements of a by now well-worn literary *topos*, the folly of navigation.[2] Thus, when Pliny contrasts agriculture and seafaring in *HN*

[1] Hor. *Od.* 1. 3. 12. See R. G. M. Nisbet and M. Hubbard, *A Commentary on Horace Odes, Book I* (Oxford, 1970), 50.

[2] The basic objections were that seafaring was an impious reversal of nature (*Terrestre animal homo*, Colum. *De Re Rust.* 1, pref. 8) and was undertaken from motives of avarice. See especially Nisbet and Hubbard (n. 1) on Hor. *Od.* 1. 3. 9–24. The idea is found in Hesiod (*Op.* 23 ff.) and in Aratus (*Phaen.* 110 ff.), Lucretius, Virgil, and Tibullus among others. A lengthy list of examples may be found in K. F. Smith, *Tibullus: The Elegies* (New York, 1913), 245–7.

14.52, we might expect the point to be that the latter is impious and irresponsible. Closer examination suggests that this is not the whole story. The context of *HN* 14.52 is one of comparative profit. Pliny quotes stories which 'prove conclusively that the merchant does not make more profit (*plus conferre mercatori*) by rashly trespassing on the sea... than is yielded by a diligently cultivated homestead (*sedulum ruris larem*)'. Pliny's views on man's activities on land and at sea are in fact influenced as much by the practical as by the ideological associations of his view of *vita*. These ideas are the subject of this chapter.

1. Terra

A. AGRICULTURE: AN IDEOLOGY OF THE SOIL?

(i) *Ideology and practicality*

Pliny's particular interest in *Terra* as man's own portion of nature does not come as a surprise. There is behind it a traditional 'ideology of the soil' that was felt to be particularly appropriate to Roman *mores* and is to be found in all Roman agricultural writers. A hard-working and moral life of rustic simplicity both did honour to and had its honour increased by the eminent statesmen of old who were personally involved in the working of their own lands.[3] The idea found favour with popularizing Stoics such as Musonius Rufus,[4] because it was in accordance with

[3] Columella, *De Re Rust.* 1, pref. 12–17; Varro, *RR* 2, pref. 3. Virgil often expresses this ideology in the *Georgics*, e.g. 2. 458–74; in 2.513–42, the simple, healthy, and peaceful life of the farmer is compared to that of the ancient Sabines, the traditional type of sturdy simplicity (e.g. Cic. *Pro Lig.* 11). The comparison with a golden age of easy plenty, however, is not entirely in accordance with his picture of the hard realities of present-day farming in *G.* 1.121 ff. Throughout the *Georgics*, however, there is a traditional ideology which R. Syme (*The Roman Revolution* (Oxford, 1939), 451) described as giving to 'moral and military excellence the primacy over pecuniary profit. If the growing of corn brought no money to the peasant, if his life was stern and laborious, so much the better.' It fits in with another ideal, the self-sufficiency of Italy and the condemnation of superfluous foreign luxuries (see sect. 2, esp. D). The actuality was different, even among small farmers: these 'may more correctly be regarded as small capitalists than as peasants' (ibid. 450). See sect. 1 A. iii.

[4] Musonius Rufus fr. 11; see A. C. van Geytenbeek, *Musonius Rufus and Greek Diatribe* (Assen, 1963), 129–30, who links these ideas with those expressed

nature to receive sustenance from the earth, which was mother and nurse to man, and also because it was a way of being self-sufficient. Columella has a typical piece of moralizing on the virtues of the farmer-generals of old at the expense of the luxury-lovers of today (*De Re Rust.* 1, pref. 13 ff.). He also suggests the divinity of *tellus*: '. . . Earth, to whose lot was assigned a divine and everlasting youth, and who is called the common mother of all things . . .' (1, pref. 2). If the yield is not good, that is the fault of man who has 'handed over agriculture which all the best of our ancestors had treated with the greatest care, to the very worst of our slaves, as though to an executioner for punishment' (1, pref. 3). Pliny, however, makes the divinity of *Terra* even more emphatic by punning on the word *coli*, 'cultivate'/'worship', in *HN* 18. 21, when he says of the use of slave-labour that 'the earth, whom we call mother and whose cultivation is spoken of as worship, is not so dull that, when we obtain even our farmwork from slaves, we can believe that this is not done against her will and to her indignation'. The personification and sympathetic feeling in this passage is typical of Pliny's descriptions of the misuse of divine nature (cf. Ch. I. 3).

It is equally clear, however, that, if Pliny's individual ideology finds more vivid expression than the traditional remarks of Columella, both authors also have practical reasons for their comments. Pliny suggests that crops prosper in the hands of *honesti* because everything is done with greater attention, *curiosius* (18. 19). This sentiment is echoed in *HN* 18. 36: 'farming done by gangs (*ergastula*) of criminal slaves is utterly bad, as is everything done by desperate men'. It is clear that Columella, too, not only dislikes *latifundia* because such vast expanses are ill-cared for by debtors and chained slave-gangs (*De Re Rust.* 1. 3. 12), but sees slaves in general as the least satisfactory workers on a farm (*De Re Rust.* 1. 7. 6–7). Concern for *cura* and *diligentia* (*HN* 18. 19) on the land are practical rather than idealistic considerations.

So far, it is clear that the agricultural writers, in particular the near-contemporaries Pliny and Columella, share some basic attitudes towards their subject. What, then, are we to make of

in Arist. *Oec.*, where agriculture is said to be natural and just ($1343^{a}26$–31) and to make its practitioners strong, resilient, and good soldiers ($1343^{b}1$–7). Cf. connections with honour and courage, Cic. *De Off.* i. 42, 150–1; Cato and Columella, prefaces.

the attempts which have sometimes been made to portray Pliny and Columella as 'diametrically opposed' on the theory and practice of farming?[5] Such a contrast makes Columella cater for 'big' landowners with large slave-gangs and capitalistic intentions, and Pliny write for the humbler farmer who has little in the way of monetary resources or intellectual capacity. Pliny's condemnation of *latifundia* (18. 35) has been interpreted in accordance with these conclusions, and he is said to be condemning the large estates concerned with mixed farming which Columella advocates. Yet we have already seen that Columella, while recognizing the use of *ergastula*, makes it clear that he prefers other forms of labour. As to the size of the farms discussed by each author, it seems that, in fact, Pliny and Columella are close to each other in both moral and practical considerations. It was clearly a commonplace of contemporary moral censure[6] to list possession of large tracts of land among luxuries and excesses, as Seneca does in *Ep*. 88. 10 and 89. 20–1. Pliny himself adopts the idea that earth is a *punctum* in the universe as a whole in order to express the madness of extending one's own land at the expense of one's neighbours' (2. 174). His famous remark '*latifundia* have ruined Italy' (18. 35) may, however, refer to a practical necessity. Moderation is needed if optimum return for the labour expended is to be achieved. He quotes a traditional maxim to sow less and plough more in the same context as the *latifundia* remark. Columella, too, urges moderation in amount of land for this reason (*De Re Rust*. 1. 3. 12); possession does not suffice if you lack the wherewithal for cultivation. Therefore it is likely that Pliny's remark is occasioned by the neglect of large tracts of land. Ranchland in the south of Apulia, where depopulation was also a problem (Cic. *Att*. 8. 3. 4), has been suggested.[7] This, moreover, is exactly the type of land Columella describes in *De Re Rust*. 1. 3. 12: men of enormous wealth possess lands of which they cannot even make the rounds, and either leave them to be trampled by cattle and wasted by wild beasts, or keep them occupied by debtors or *ergastula*.[8]

[5] R. Martin, *Recherches sur les agronomes latins* (Paris, 1971), 377–8, 393.

[6] A. Cossarini, 'Latifundia perdidere Italiam', in *Plinio il Vecchio*, 253–4.

[7] K. D. White, 'Latifundia', *BICS* 14 (1967), 62 ff.

[8] P. A. Brunt (review of K. D. White, *Roman Farming*, *JRS* 62 (1972), 154 ff.) points out that 'large' landowners might in fact have a number of smaller holdings, so as to have more security against weather risks (Pliny, *Ep*. 3. 19. 4) or

Pliny's ideas on practical farming, then, have much in common with those of Columella. We may note in conclusion that even the great respect Pliny feels for that embodiment of Roman tradition, the Elder Cato,[9] does not cause him to idealize the past to the detriment of present-day agriculture. Columella denied the idea that latter-day decline in yield indicated an exhaustion of the earth. He prescribed greater care and industry on the part of the cultivators (*De Re Rust.* 1, pref. 3 ff.). Pliny can see a positive improvement in certain areas: there has been a great increase in the number of different vine-species cultivated in Italy since the days of Cato (14. 45). This suggests that he, too, was not interested in the 'soil decline' theory. Equally, it suggests that, in his view, the present-day farmer was not necessarily lagging behind his predecessors in *cura* and *diligentia*. Pliny is, as usual, positive as well as practical. The Virgilian image of the poor peasant farmer scraping a living off a land overcast with regret for a long past Golden Age of plenty[10] is remote from this attitude. Closer examination of the type of farmer to whom Pliny addressed himself and the attitude shown towards labour in agriculture should dispel it completely.

(ii) *Landowner and peasant farmer in the* HN

Examination of the *HN* soon makes it clear that Pliny is not concerned exclusively with a small peasant farmer on the Virgilian model. In sect. 1. B we shall see that any special interest in the smaller farmer is not due to the latter's representing some kind of moral ideal. In any case, the terms 'large' and 'small' are in many respects misleading and need not imply antithetical approaches to farming. Many of Pliny's agricultural preoccupations apply equally to all farmers.

To start with, such an antithesis suggests that Pliny was not

disease (Varro, *De Re Rust*, 2. 3. 9). Pliny (18. 35) says Pompey never bought land belonging to a neighbour's estate.

[9] 'M. Cato, whose authority is very little enhanced by his triumph and censorship; so much more comes from his personality' (29. 13). Here, Pliny fully supports Cato's opinion of Greek doctors (see intro. 19). However, he is realistic enough to recognize that Cato is not an omniscient oracle; he could not have foreseen present-day abuses in the art of medicine. Again, while laying great store by Cato's agricultural precepts (18. 25–6), Pliny will not follow them mechanically; the situation would differ according to location (18. 30).

[10] *G.* 1. 121 ff.; see n. 3.

concerned with the interests of the society of landowners of which he was himself a part.[11] Disapproval of luxuries of the table might prohibit any mention of game-preserves found on such men's estates (17. 50). However, he was well acquainted with, and not necessarily opposed to, the aesthetic aspect of these estates, as has been suggested in the discussion of his passages on gardens.[12] There are also suggestions of a contemporary realism with regard to the rival claims made on the time and attention of influential Roman landowners. Pliny seems more content than Columella or Cato to leave the farm in the hands of a reliable bailiff.[13] He criticizes Mago's idea that, upon buying a farm, one ought to sell one's town house: such an opinion was too rigorous and against the interests of public affairs (18. 35). Pliny recognizes very well the claims of political service upon wealthy landowners. 'Too rigorous' also suggests that, at least for those to whom farming was not a necessity, it should not be made an uncomfortable obligation either.[14]. Appreciation of the gentler aspects of affluent country life influences Pliny's attitude to farm work in several ways.

(iii) *Work: the practicality of labour-saving*

For the affluent landowner, the concept of *labor* as envisaged by Virgil, the hard physical struggle of man in direct contact with the soil, was of little relevance. The incessant round of ploughing

[11] M. Frederiksen, 'Plinio il Vecchio e l'agricoltura in età imperiale romana: gli aspetti tecnici ed economici' in *Tecnologia*, 81–97.

[12] Ch. II. 5; Pliny, *Ep*. 5. 6 for a detailed description of the beauties of a country estate.

[13] Owner supervision: Pliny considers that the bailiff, besides the qualities mentioned by Cato, *De Agr*. 5. 1–5, should possess intelligence nearly equal to that of his master, though he should not himself realize it (18. 36). In 18. 35, he implies that he approves of regular visits, at least, from the owner. Mago was overzealous, 'although this showed he laid stress on constant overseeing'. However, the overall impression is that there is less emphasis on the owner and more on the bailiff. In contrast, Columella is much more emphatic about the need for regular personal supervision by the owner of both farm and bailiff (*De Re Rust*. 2. 1. 18–2. 2). Cato, too, though he takes it or granted that the owner will visit rather than reside in the country, gives him a very full list of duties to carry out when he does go there (*De Agr*. 2).

[14] See Ch. II, intro., cf. Brunt (n. 8), 154, on an actual reluctance on the part of the upper classes, especially those in Rome, to be involved in agriculture, even though this did not, in their case, mean manual work. More and more become absentee landlords like Pliny's nephew.

and sowing called at most for indirect participation on his part, the overseeing of such operations by the farm's workforce. He would, however, be anxious to save the labour of his employees if this made his estate more cost-effective.[15] Pliny's interest in certain technological innovations in the sphere of agriculture suggests an affinity with this viewpoint. Indeed, he is more concerned with such inventions than is Columella. This apparent discrepancy in the traditional contrast between Pliny and Columella has been explained by the suggestion that Columella's type of farming did not need to cut the cost of large-scale slave labour, whereas Pliny is catering for those to whom this option is closed.[16] It may well be that Pliny was made aware of these innovations in Gaul, where there seems to have been less slavery and a greater emphasis on a quasi-feudal peasantry in farming.[17] Pliny has half a dozen references to Gallic techniques, including the wheeled plough (18. 171–3), a larger type of scythe (18. 261), and a harvester (18. 296). However, it is also clear that such techniques are for large-scale farming and would probably be of comparatively little use to a small-scale farmer. In both *HN* 18. 261 and 18. 296, Pliny describes the Gallic estates as *latifundia*, with no trace of disapprobation. The possibility that less slave labour was used in Gaul might make him more favourably disposed, given his practical and ideological objections to *ergastula*. It is more likely that the use of the innovations he describes made such estates more efficient to run.[18] Hence, there would not be the neglect postulated as the reason for his criticism of the Italian *latifundia*.

[15] Cost-effectiveness was a concern of the agricultural writers. Varro says that the farmer should grow his own vine-trellises to ensure that the vineyard does not eat up the profits (*De Re Rust*. 1. 8. 2). Columella says stone-clearing should only be done if there is cheap labour available (*De Re Rust*. 2. 2. 12), and that the cost of heavier draught animals is outweighed by the profit from the improved crops (*De Re Rust*. 2. 2. 24). See Brunt (n. 8), 154.

[16] For Martin (n. 5), who blames a Roman lack of interest on the existence of the slave economy (73–80), the 'reactionary' Pliny's interest in technological innovation is a paradox (393) which he cannot explain.

[17] See, most recently, J. E. Drinkwater, *Roman Gaul* (London, 1985), 161 ff.: it is generally agreed that the Gallic estates were run on the labour of an at least nominally free peasantry (171). Cf. E. M. Wightman, *Gallia Belgica* (London, 1983), 129.

[18] The precise motives for such technical improvements as the harvester are uncertain. Shortage of labour is suggested, as is the need to get the harvest in as quickly as possible in case of bad weather: Wightman (n. 17), 121–4. In any case, greater efficiency seems to be the general motive. As Wightman suggests, the

In all his examples, Pliny emphasizes the saving of labour. The wheeled plough means that a field does not require hoeing (18. 173), but the method does require two or three pairs of oxen. More specifically, he says in *HN* 18. 261 that the larger Gallic scythes economize on labour by cutting through the stalks of the grass in the middle and missing the shorter ones. Labour-saving is implicit in the use of the harvester described in *HN* 18. 296, since this is drawn by a team of oxen, whereas hand-picking is the method employed elsewhere. The interest of the *dominus* in such labour-saving and cost-effective farming generally is suggested by specific references to him in other passages. In *HN* 17. 191–2, Pliny warns large-scale farmers not to trim their vines too early in spring in an attempt to save labour, in case there is a late frost. Again, to be too painstakingly productive is not economically viable: 'there are some crops it does not pay the owner to harvest if the cost of the labour is reckoned up' (18. 38).

Pliny's interest in the concerns of the large-scale farmer seems indisputable. However, labour-saving as a general practical principle should by no means be excluded from the interests of those who farmed on a smaller scale. It would be difficult to believe that these did not wish to maximize gain and minimize their personal labour. Some of Pliny's remarks on labour-saving have a more general application; his comment on the planting of trees puts emphasis on the sheer laboriousness (*operose praecipi arbitror*, 17. 75) of the more painstaking method of intermediate transplanting, even if it does produce larger trees. It is pertinent to consider whether Pliny's concern for the alleviation of *labor* goes beyond the practicalities so far discussed.

(iv) *Work: moderation in Pliny's philosophy*

In *HN* 17. 209–10, Pliny suggests that rest is, like day and night, an integral part of the harmony of all nature: even a vine should be released from its ties for a few days, 'for just as beasts of burden released from the plough and hounds after a chase like a roll on the ground, so even the vine's loins like a stretch. The supporting tree, too, enjoys being relieved of its continual burden, like a man getting his breath back. There is nothing in

stimulus may well have been the prospect of good profit and heavy yields, which overcame the disadvantage of the loss of the straw with such a machine.

Nature's handiwork which does not desire regular rest periods after the pattern of days and nights.' It may be difficult to regard this picturesque intrusion into a factual and practical account of viticulture as being entirely serious, but the core of the idea should not be overlooked as it appears again in *HN* 18. 178 with regard to the rest days for farm animals: 'Nature has appointed laws even for the labour of animals.' Man, by implication, is even more entitled to them. It is, of course, true that Pliny quotes ancient precepts concerning work and making optimum use of one's time (18. 40; 18. 43: 'agriculture depends on expenditure of labour and this is the reason for the saying of our ancestors that on a farm the best fertilizer is the master's eye'). His own attitude, however, may well be analogous to his views on fear (Ch. II. 3). Work, like fear, should not become something that threatens to engulf man physically and mentally, creating pessimism and anxiety. The key is to be in control of the work, not the other way round. Pliny is aware that man has a great capacity for work; he can, if necessary, plough difficult ground without oxen. 'Man has such a capacity for labour that he can actually perform the function of oxen' (18. 178). It is the right and natural way of achievement. He quotes with approval (18. 42–3) an old story about a successful farmer, C. Furius Chresimus. He, like Acilius Sthenelus, who also produced remarkable agricultural results in Pliny's own time (sect, B. i), was a freedman. When accused of using magic to entice away the crops of his neighbours, he produced his tools in court, saying 'these are my magic spells, citizens, and I am not able to exhibit to you my midnight labours, early risings, and sweat and toil'.[19] But the life of drudgery, like the life of fear, denies man's rational supremacy. There is contemporary evidence for this in popular Stoic thought: Musonius Rufus suggested that certain farming tasks, such as shepherding, were preferable to others, which in his view required more unremitting physical labour,[20] because they left the mind freer. As we shall see, Pliny suggests that

[19] R. Garosi, 'Indagine sulla formazione del concetto di magia nella cultura romana', in P. Xella (ed.), *Magia: Studi di storia delle religioni in memoria di Raffaela Garosi* (Rome, 1976), 33–5, 63, saw the story as contrasting Roman *virtus* and *lex* with magical arts. It could equally well illustrate that *ratio* and *labor*, man's natural use of his mental and physical potential, prove that there is no need for recourse to irrational and unnatural solutions (see Ch. III. 2. A. ii).

[20] Musonius Rufus fr. 11.

rationality should be applied to the problem of *labor*: the key to farming is the intelligent organization of work, the mastery of theoretical as well as of practical skills.

B. *ARS AGRICULTURAE*: *RATIO* IN FARMING

(i) *Theory and practice*

A contrast between Pliny and Columella has sometimes been drawn on the grounds that Pliny disliked the theoretical knowledge that Columella advocates.[21] We have already seen that this is not an accurate representation of Pliny's views generally (Ch. II. 2). It can, in any case, be shown that Columella was himself in favour of a blend of theoretical and practical knowledge, *ratio* and *experientia*.[22] This is very similar to Pliny's ideas on *ratio*, *usus*, and *observatio*. Indeed, Pliny's conviction that learning as well as practical experience is a vital ingredient in agriculture is illustrated not merely by brief phrases but by longer anecdotes. Two of these deserve closer attention.

The first concerns the success in viticulture of the grammarian Remmius Palaemon. He employed a freedman's son, Acilius Sthenelus, who had already won renown in this field. Eventually he produced astonishing vintages from a farm in a poor and neglected area. The estate was then bought from him at four times the original price by another intellectual, Seneca, even though Palaemon was an enemy of his. Of Palaemon, Pliny says that 'he undertook the cultivation of the property not from any high motive (*virtutis animi*) but at first out of vanity, for which he was notorious. He had the vineyards redug and retrenched under the guidance of Sthenelus and got the estate into an almost incredibly marvellous condition, though he himself was only pretending to be a farmer' (14. 50). Did we not know of Palaemon's unpleasant reputation from another source (Suet. *De Gramm.* 23), which accords with Pliny's sarcasm about his motives, we

[21] Frederiksen (n. 11). See Ch. II. 2.

[22] Brunt (n. 8), 153–4. Columella, *De Re Rust.* 1, pref. 4, 'agriculture is without doubt most closely related and as it were own sister to wisdom'. It must be tested by experience: 1. 1. 16 '*usus* and *experientia* hold supremacy in arts', and above all 'a method (*ratio*) recalled from times past must be fortified by many examples' (3. 7. 3).

might suspect that the 'fake farmer' remark was directed against *litterati* who theorized about agriculture while others like Sthenelus had the practical ability. The real irony of Palaemon's learning is, however, expressed later, in Pliny's description of his neighbours' reaction. 'The lazy neighbourhood vindicated itself by attributing his success to his exceptionally profound studies' (14. 51). The point is that they were too lazy to see that Palaemon's learning was not in this case too abstruse: in employing and imitating Sthenelus' methods he had made a very pracical use of his intelligence which was not, in reality, 'beyond' them. The educated *dominus* is important in practical farming. He recognizes where the expertise lies even if he does not participate in the physical labour.[23] Pliny also thinks it worth adding that the next purchaser, Seneca, was 'the most learned man of his day' (14. 51).

Another defence of the place of the intellectual in farming concerns the astronomer Democritus (18. 273–4). His learned interests were spurned by his fellow citizens. However, he realized the bad effect the weather at the time of the rising of the Pleiades had on the olive-crop. This theory is clearly of importance in Pliny's eyes since he refers back to a previous mention of it and announces a fuller exposition.[24] Democritus, to prove his point, bought up olive-oil while it was still cheap, 'to the amazement of those who knew that the things he most prized were poverty and learned repose'. When he had made his point and become rich, he gave the money back to the now repentant landlords. A similar demonstration was later made by the Roman philosopher Sextius Niger. Pliny concludes that 'such is the opportunity offered by learning which it is my intention to blend into my account of agriculture as clearly and convincingly

[23] Suet. *De Gramm.* 23, 'he cultivated his fields with such care that it is common talk that a vine grafted by his hand (*manu eius*) yielded 300 bunches', suggests that Palaemon's interest did extend to some personal participation.

[24] *HN* 17. 11: the Pleiades rise at the end of spring; the rain which occurs at this time is bad for olives and vines because this is their fertilization season. Pliny stresses that it is, therefore, a critical time (*hoc est illud quadriduum . . . decretorium, hic articulus . . .*). Compare 18. 280, where Pliny says that their rising marks the beginning of summer, their setting that of winter, embracing in the six months' space between them the harvest and vintage and ripening of all vegetation. It is by giving him glow-worms, a rural equivalent of the Pleiades, that Nature has shown *benignitas* to the simple farmer who may not be able to understand astronomical theories (see B. I, *fin*).

(*dilucide atque perspicue*) as I can'. Pliny, then, is in favour of the use of learning in agriculture. However, it must be blended with practical skills and it must not be too obscure. In the cases of Palaemon and Democritus, their strength is not in their learning *per se*—though Democritus' is of more direct relevance than is Palaemon's—but in a more general application of intelligence to practical purposes. Palaemon's neighbours were too sluggish mentally as well as physically to appreciate this difference. The story of Democritus is a reproof to *domini* who think that wealth alone creates success and spurn knowledge as being useless for advancement in any practical sense.

The astronomical theory studied by Democritus was closely allied to the general understanding of nature, and its benefits in this respect are praised by Pliny (Ch. I. 4; II. 3). However, it was not an easy art to master. This is reflected in Columella's introductory comments on the knowledge needed to farm successfully. He too can represent such knowledge as a general understanding of nature: the aspiring farmer should be 'rerum naturae sagacissimus' (*De Re Rust.* 1, pref. 22). But he soon goes on to emphasize its technicality, including various astronomical and meteorological elements 'which I cannot believe that any man can know beforehand without the light of intelligence and without the most expert instruction' (1, pref. 23). Where Pliny differs from him is over the emphasis to be put on the difficulty and exclusiveness of the theoretical knowledge. There is a conscious attempt in the *HN* to make such knowledge accessible to a wider audience, to explain it *dilucide* in order that more people might have a better chance of understanding it. This is in keeping with Pliny's philosophical principle of universality, where the emphasis is on mankind as a body. This idea extends to his vision of the *HN* itself as an all-embracing work, *vita/natura* in all its forms (pref. 13, 15). The significance of this literary attitude as an attempt to rival Columella personally cannot be ignored and will be examined in the next section. The attempt to make theoretical knowledge more widely appreciated is fraught with difficulties as the following examples show.

In *HN* 18. 204–6, Pliny lays out rival ideas on sowing. 'There are those who hurry and lay down that, while sowing in haste sometimes proves disappointing, sowing late always does. Others on the opposite side think that sowing even in spring is preferable to sowing in a bad autumn . . . Some ignore meteorological niceties and fix limits by the calendar . . . These latter pay no attention to Nature, while the former pay too much, and this gives rise to blind theorizing, since the matter is disputed among rustics who are ignorant of literature as well as of astronomy.'

Pliny thus outlines the problems. The implication is that a mean should be found. The rustics must learn to use knowledge more subtly, the learned more practically. The achievement of the latter was illustrated by the stories discussed earlier. In *HN* 18.206, Pliny attempts to tackle the rustics: 'It is an arduous and vast aspiration to join the divine science of the heavens to the ignorance of the rustics.' Here, he comes close to admitting the truth of Columella's 'exclusive' view of such knowledge. It becomes clear by § 208 that the secret of practical application of such theories lies in recognizing the unpredictable, the behaviour of certain stars, hailstorms, and rain, which disturb the simplistic interpretations of the celestial rules: 'they disturb the regularity of our expectations' (208). Or, as he puts it in *HN* 18.238, the farmer has to start his spring tasks not when the west wind ought to blow, but when it does blow.

It is this difficulty which leads Pliny on occasion to abandon the stars as a guide and to recommend to 'the simple farmer unlearned in astronomy' the observation of the countryside immediately around him. This furnishes evidence of the larger-scale cosmic changes, but can have the advantage of being particular as well as universal. At least, that is what Pliny seems to be saying when he draws attention to the significance of the falling of the leaves. 'In this way the year's weather may be estimated, as they fall later in some places, sooner in others; for the weather is perceived by the nature of the climate and the locality; and this method contains the advantage that, while it is universal and world-wide, it is also peculiar to each individual locality' (18.226, cf. signs of the season given by the leaves 18.267 ff.). Such signs are recounted even more enthusiastically in 18.251–3: *Natura* in her *benignitas* has provided the rustic with glow-worms to mark the change of season, and with plants to mark the time of day. In this way, she tells him (252), she has dispensed him from the need to look at the sun or Pleiades for guidance. However, Pliny admits in 18.209 ff. that some signs of this type are not foolproof. The appearance of migratory birds and butterflies does not always indicate that the cold weather is at an end. Hence, as Virgil saw, some knowledge of astronomy is needed as well, in order that the movements of the planets, which appear to be responsible for the irregularities of the weather Pliny has been describing, may be watched and understood.[25]

Pliny is at pains to stress that there can be an understanding of nature at more than one intellectual level. In certain cases, *observatio* alone, with little or no theoretical knowledge, will suffice the *agricola indocilis* for all practical purposes. A cosmic sympathy causes the less significant parts of nature to reflect the behaviour of the celestial bodies; glow-worms are the countryman's guiding constellations. 'See, you have the Pleiades at your feet,' *Natura* tells the farmer. 'It is certain that they are the offspring of this particular constellation', adds Pliny (18. 253). At the beginning of book 19, he congratulates himself for providing an account of the constellations, seasons, and weather which is easy even for non-experts to understand. He adds that he has also shown that 'for those who really understand the matter, the countryside contributes to our knowledge of the heavens no less than astronomy contributes to agriculture'. This expresses the idea of sympathy once more. In many cases, country signs are as true an indication of *Natura*'s changes as are the stars. However, it also implies that those who 'truly' understand have a knowledge of both. In this way, those irregularities unforeseen even by plants and animals may be taken into account, if not actually predicted. It is this sort of wider understanding which is necessary in Pliny's eyes for successful farming. Mechanical *observatio* alone will not make a good farmer, as the stories about Palaemon and Democritus prove.

(ii) *Literary rivalry in Pliny's account of agriculture*

Pliny's interest in instructing the less educated in agriculture owes much to his general philosophy of the need to benefit mankind as a whole. But this is not the whole story. His pointed inclusion

[25] See 2. 105–16 for the complexities of the effects of the stars upon the weather. Pliny says that it is not just the planets which have an effect. Other stars may do so, as may constellations (106) and parts of the constellations (108). Other weather changes are less predictable and are not guided by the stars (111–16). For regular winds, see also Ch. I. 4.

of references to the *indocilis agricola* in what was, after all, a literary work unlikely to be read by such a class suggests a further motive. The status of the *HN* as a literary work accounts for some of this emphasis and also provides the most likely explanation for those passages which have been commonly regarded as attacks on Columella. Pliny has had to justify his account against the other agricultural writers. It has been suggested that he is eager to improve on Virgil at times.[26] When a contemporary like Columella writes an agricultural treatise in the more utilitarian 'handbook' form, the rivalry becomes even more acute. This explains why Pliny makes sharp criticism of Columella on points which do not seem to merit such a reaction on purely factual grounds.[27] There is also a literary as well as a philosophical reason for proclaiming his work to be of universal proportions, embracing the whole of nature. Columella had emphasized the magnitude and diversity of the field of agriculture alone, '. . . so varied and so vast a science' (*De Re Rust.* 1, pref. 22 ff., especially 28). It is in Pliny's interest to suggest that other writers are not as useful as he is. Virgil's assignation of agricultural operations to numbered days of the moon is seen as too arbitrary in its subtlety, and perhaps therefore liable to misuse by the ignorant rustics of *HN* 18. 205–6 (above, B. I). Pliny instead advocates the practical value of general rules, *legum utilitas* (18. 321). It is convenient to be able to suggest that your rivals suffer from excess of minutiae, that they run the risk, in effect, of narrowing their subject's appeal to an intellectual élite and depriving it of *utilitas vitae*. Pliny tried to explain astronomy to the ignorant

[26] R. T. Bruère, 'Pliny and Virgil', *C. Phil.* 51 (1956), 228–46, argues that Pliny tries to correct Virgil as often as possible, despite expressions of admiration for him. However, he goes too far in suggesting anything more than an occasional assertion of being able to go one better than a revered predecessor.

[27] For Pliny as critical of Columella, see Martin (n. 5), 376 ff. But only one out of the four definite references to Columella criticizes him for an actual mistake. Other evidence is on the unsure grounds of omission and allusion. Rather than concluding that Pliny was entirely opposed to the whole of Columella's outlook, a passage such as 18. 70, where Pliny comes down severely on a trivial particularity, suggests a more personal animus. Literary rivalry with a contemporary agricultural writer rather than criticism of him as an agriculturalist *per se* is the more appropriate conclusion to be drawn from 19. 2. 'We are surprised that some persons seeking from these subjects *the satisfaction of knowledge and a reputation for learning*, have passed over so many matters. . .'. If it is really an allusion to Columella, the wording expresses surprise at his assertion of literary and scholarly eminence.

rustic because of 'the vast benefit it bestows on life' (18. 206). By anticipating the rise in olive-prices which would follow the stormy weather at the rising of the Pleiades, Democritus gave a practical demonstration of how to use learning to a wider, less informed audience. As a Plinian example of 'service to mankind', we may compare it to the services of the explainers of eclipses, who freed the less wise from fear (2. 54, Ch. II. 3).

It is possible that Pliny felt inferior to Columella on stylistic grounds, but it is clear that literary self-consciousness in general affects him more in the agricultural sphere than anywhere else. His remarks made in the context of literary apologia about his subject as being barren material, life in its least elevated aspect, and the consequent need to use uncouth rustic vocabulary (pref. 13) have often been noted.[28] The opening paragraphs of book 18 tend to escape notice. Here the justification of literary reputation and theme are once more bound together, but the 'rustic' apologia is more individual. A defence of earth against the misuse and ingratitude of man is skilfully combined with a defence of the down-to-earth subject-matter of the *HN* against spiteful critics. These are compared in their negative and unproductive criticism to 'bad' products of nature, animals of evil omen and poisons (18. 4–5). Like all evils of nature, however, they are a tiny minority. Pliny allies himself with *Natura* and *vita* against these critics, the 'brambles' of the human race (18. 5, cf. 17. 96, Ch. V, intro.). This leads him to a justification of his rural theme, one that may seem lowly but in fact is honourable as it is of supreme use in 'life'.

The appearance of such an elaborate digression half-way through the *HN*, at the beginning of the main part of the discourse on agriculture, marks this subject out as the core of Pliny's philosophy, the epitome of man's life as lived in harmony with *Natura*. The existence of a formidable literary rival in this very field gives personal urgency to the traditional forms of apologia.

(iii) *Pliny's ideal farmer: a summary*

Taking stock of all Pliny's views on agricultural theory and practice, we can try to depict his ideal farmer. The first part of the

[28] e.g. G. Pastucci, 'La lettera prefatoria di Plinio alla *N.H.*,' in *Plinio il Vecchio*, 184–5.

discussion suggests that he belongs to a reasonably affluent and educated stratum of Roman society. He is not *indocilis* but not necessarily *litteratus* either. The second part suggests that he must be intelligent enough to take a wide, overall view, rather than keep narrowly to given rules. That is, he must have a reasonable amount of knowledge, never too abstruse, literally at his finger-tips; he must be capable of blending it with practical abilities and of modifying his theoretical approach in accordance with particular situations: localized or unforeseen variations on general patterns.

These qualities are similar to those required of the military general. The cast of mind capable of the large-scale organization of both long- and short-term strategy would be well fitted to take on *ars agraria* as well.[29] The first hint that this is indeed in Pliny's mind occurs in *HN* 18. 19 ff. We saw that the farmer-generals of old were praised not just on moral grounds but also on those of practical skill: 'They treated the seed with the same care as they managed their wars and marked out their fields with the same attention as they gave their camps.' Pliny gives a practical and contemporary significance to the old Roman picture of military and agricultural *virtus*. This is different from the more traditional passage in Columella (*De Re Rust.* 1, pref. 17–19). The hard work of country life as opposed to urban idleness and luxury ensured that the Romans easily endured the hardships of military life. For Pliny, the idea is reversed and the mental as well as physical qualities of the general also stand him in good stead in agriculture.

This idea is reinforced by the anecdote in *HN* 18. 32. Pliny discusses the size and situation of the farm-house in relation to the farm. He, like Columella, mentions the mistakes of proportion made by L. Lucullus and Q. Scaevola (Colum. *De Re Rust.* 1. 4. 6). But, whereas Columella leaves specific examples at this point, Pliny adds another one, the general C. Marius, to illustrate the ideal. 'Correct arrangement requires some technical skill. Last of all, C. Marius, who was seven times consul, built a house in the district of Misenum, but he used the practical skill (*peritia*) he had acquired from laying out camps so that even his rival Sulla declared that the others had been blind compared with Marius.'

[29] The suitability of the military mind: see Frederiksen (n. 11), and cf. Tarius Rufus, a distinguished ex-soldier, below and n. 30.

The blend of intelligence and technical skill needed is indicated by *peritia*. Its correctness is shown by the remark of Sulla. Marius was not *caecus*, either in overtheorizing like the literary experts in *HN* 18.205 (*caeca subtilitas*), or in sneering at the propounders of useful theories in the manner of the critics of Democritus. He was practical in his approach as well, but with a greater understanding and depth of vision than the unlearned farmer relying unthinkingly on his 'laws'. The layout of the farm requires *observatio* of the surrounding countryside (18.33) but wider organizational and conceptual powers need to be applied to the results. Despite Pliny's attempts to widen the scope of his farming instruction, the rustic equivalent of the common soldier is not going to exercise the skill of the farmer-general. If Pliny's ideal farmer existed anywhere, he would most probably be a man similar in position, education, and experience to himself. That is, he would not be an aristocrat, but a man of equestrian rank, the sort whose practical skills had played an important role in the imperial service. Perhaps we should add, more generally, men who had advanced themselves from obscure origins by the energetic exploitation of such abilities: the over zealous Tarius Rufus,[30] for instance, who 'though of extremely humble birth, won a consulship through military efficiency' (18.37), or the enterprising freedman like Acilius Sthenelus and Furius Chresimus discussed earlier.

It is certainly reasonable to conjecture that Pliny's ideal of the farmer-general owed a great deal to his own background, with its distinguished military career. Such a background would add *auctoritas*, both literary and practical, to his repute as an expert on agriculture.

2. Mare

A. *MARIA TEMERATA*: IMPIETY TOWARDS NATURE?

The sea, as we have seen, was the element which embodied nature's wild side. At best, it was *incerta*, at worst, *saevissima*,

[30] L. Tarius Rufus is adduced by Pliny as an example of excessive *cura* and *diligentia* (18.37). He farmed so well (cf. *optime* in a paradoxical sense, 36) that his heir refused to take over the estate. The farmer must also be aware that it is not always profitable to harvest certain crops or lavish attention on certain areas of land.

even *bruta Natura* with no discernable *ratio*. But is the sea so hostile to man's encroachment? Does *maria temerata* in *HN* 14.52 (see Ch. V, intro.) imply a notion that had long been a commonplace in ancient literature, that such attempts are even impious?

We saw that the sea is naturally in conflict with the land in the *HN*. But Pliny does not seem to view it as being against man personally, unlike Lucretius, who describes it as being untrustworthy because of its stealth, violence, and cunning (2.557). This hint of personal malevolence appears again in Lucretius' poem (5.999–1005), where the mass slaughter of men by the sea is mentioned in the same passage as, but distinct from, the slaughter of men by each other in battle.

Pliny in contrast views the savagery of the sea as that of one natural element directed against another. There is a continual struggle between *Mare* and man's element *Terra*, as we saw earlier, but the idea of the sea defrauding man personally, as in Lucretius, is rare in Pliny. In *HN* 3.81, he describes a sea-level island, not the sea itself, as being 'treacherous to ships'. Malevolent stealth on the sea's part is suggested by *HN* 2.205, 'coasts which treacherously encroach on the land', but this too refers to the relationship between seas and land. In general, the difference in stress between the two writers on the malevolent aspect of the sea[31] corresponds to their respective views of the relative *benignitas* towards man of *Natura* generally. Lucretius is much more pessimistic in this respect than is Pliny.[32]

The attitude more apparent in Pliny is that the sea is *incertum* and man must show respect for this dangerous part of *divina Natura*, not by cowering away in fear, but by exercising care and intelligence in his seafaring efforts. The following passages provide elucidation of this attitude.

HN 19.3–6 is the most traditionally moralizing of all Pliny's seafaring passages. The growing of flax is described as contrary to nature in that it scorches the soil. Its use in sails allows men to

[31] Malevolence; see Prop. 3.7.37, 'treacherous nature has made the sea a path for the avaricious'. K. F. Smith (n.2), 247 notes the proverbial Roman use of *Neptuni filius* to denote a 'liar, a perjurer, and a man of wrath', as in Cic. *ND* 1.63 (quoting Lucilius).

[32] Lucr. 5.196–234. See further Ch. I.3 and C. D. N. Costa (ed. and comm.), *Seneca: Medea* (Oxford, 1973), 100.

find new forms of danger and death by being borne along not only by the waves but also by the winds, which can be dangerous to man and his livelihood even on land. However, this passage is immediately preceded by enthusiastic descriptions of the wonder of communication facilitated by this plant. The main object of the criticism is not so much navigation as the flax plant itself, which unnaturally scorches earth as it grows and 'captures the winds' (19. 4–5) when processed by man. With regard to seafaring as such, Pliny's real objection is to unnecessary risk-taking. Adding more and bigger sails to ships is tempting the winds too much (*provocari mortem*).[33] A more modest approach is becoming to men when they are in contact with the wilder side of nature. Unnecessary participation in a dangerous occupation may be the point of the passage immediately following this one, *HN* 19. 7. The rich seaborne spice trade gained by Egypt might justify her maritime interests, but the same excuse cannot be offered for those of the remoter inland tribes of Gaul. Here again, however, Pliny is actually offering and denying excuses for the growing of the ambiguous flax plant which does damage to the earth (*terrae iniuriam facit*, 19. 7).

Other condemnatory passages more clearly have as their real target man's unnecessary risk-taking while navigating. Despite the unpredictable gusts (*flatus*) which are a hazard to the sailor (19. 6; Ch. V, intro.), Pliny also talks (2. 116) of the law of nature which at least gives regular winds their limits and seasons. Thus, in sailing for gain in winter (2. 125), man is not only failing to do justice to his own *ratio* but is also ignoring that of *Natura*, which prescribes the spring for sailing. '(There is) a law of nature that even winds have their prescribed limits and seasons. Accordingly, spring opens the sea to voyagers' (2. 122). That navigation is not in itself impious is implicit in this statement that *Natura* actually prescribes a comparatively safe sailing period. Man's recklessness is again Pliny's concern in *HN* 12. 87.[34] He expresses surprise that one of the foreign peoples involved in the spice-trade sets

[33] Cf. Sen. *Medea* 318 ff., which suggests that caution is the real way to rule the waves. There is a contrast between prudent handling of the sails (323–5) and the 'over-eager' (*avidus nimium navita*) sailor who runs before the wind with full sail. See Costa (n. 32), 102.

[34] A problematic passage. Pliny appears to mean the Trogodytae, neighbours of the Ethiopians, but the direction of the prevailing winds he then describes better suits traders travelling in the opposite direction, from India or China.

sail without any of the usual aids to navigation which can help man against the unpredictable blows of nature. They also sail in winter, though in this case the easterly monsoon winds which prevail at that season are an advantage. 'They convey their cargo over the vast seas in rafts, which are neither steered by rudders nor propelled by oars or drawn by sails or assisted by any rational device (*ratio*); in those regions, only man and man's boldness (*audacia*) stand in place of all these things. Moreover, they choose the winter sea . . . ' Pliny implies that, whatever the motive, it is not worth it; the round journey is said to take five years and many of them perish on the voyage.

B. *LUCRUM* OR *SCIENTIA*: THE ACCEPTABILITY OF MAN'S MOTIVES

Recklessness rather than impiety appears to be Pliny's cause for concern over navigation. What of the other objection of ancient times, cupidity? Doubts have already been cast on Pliny's adherence to this at the end of the introduction to this chapter, where it was suggested that, in *HN* 14. 52, Pliny is talking about the comparative profit rather than propriety of viticulture as opposed to seafaring. Pliny puts a good deal of emphasis on this aspect of the comparison. 'These examples conclusively prove that the merchant does not *obtain more profit* by rashly trespassing on the seas, nor by going as far as the Red Sea or the Indian Ocean to seek for merchandise, than is yielded by a diligently cultivated homestead.' Indeed, it sounds as if he has a controversial case to plead. This is borne out by the fact that elsewhere (6. 101) he gives details of the astonishing profits which could be made from the Eastern trade. It is his evidence as much as that of other sources, which bears witness to the significant expansion of such commerce in the first century. On the other side, Pliny's assertion may be supported by a contemporary development of viticulture into a highly profitable concern.[35]

Roman history showed an increasing interest in the sea as a source of profit. By Pliny's day, the fortunes of the rich and powerful often originated in trade, even if it was then trans-

[35] For the theory that viticulture had developed recently into a highly profitable concern subject to moral criticisms on this score, see N. Purcell, 'Wine and Wealth in Ancient Italy', *JRS* 75 (1985), 1 ff.

formed into land. Land was still the form of wealth that was considered to be the most reputable—*honestum*—despite the changes in the composition of the ruling classes. The two sources of income were often paired (Hor. *Od*. 1.31; Dio Chrys. 34.34).[36] Petronius, writing under Nero, has his *nouveau riche* upstart, Trimalchio, make a fortune from trading by sea, which he uses to increase his agricultural interests (*Satyricon* 76–7). Years earlier, even the admired upholder of Roman tradition, Cato the Elder, could put financial interests first and engage in 'the most disreputable form of speculation', the underwriting of ships (Plutarch, *Cato Mai*. 21.5). The comment on disreputability is Plutarch's; Cato himself, Pliny's source for one of the bumper vintages quoted just before the trade–viticulture comparison in *HN* 14.52, 'came to regard agriculture as a pastime rather than as a source of income'. He preferred to exploit other branches of *res rusticae*, pasture-land and ponds. The picture of a hard-headed businessman rather than moralist is endorsed by Cato's own words in *De Agr*. 1.1–3. Compared with agriculture, usury is not *honestum*. Trade is *periculosum*, the stricture being solely practical, not moral. In any case, Cato did not allow the risk to stop him. Plutarch says he was very cautious and managed to cut the risks down to a minimum. Petronius' hero was not put off by an initial disaster. A subsequent successful voyage more than outweighed the previous losses (*Sat*. 77).

The advantages Cato lists for agriculture in the same passage of *De Agr*. (1.4) do, it is true, contain a certain moral element: it produces the bravest men. This, together with the justification that it is the 'most stable and least invidious' occupation may suggest a general Roman anxiety. In J. H. Thiel's words,[37] 'a certain fear of political radicalism may have been a strong motive in Roman history to limit as much as possible the civic element in naval service'. Naval recruits were poor—too poor to afford the armour required for the army—and landless. Thus the men who went to sea could be portrayed as mirroring the instability and unreliability of that element, in contrast to the men who worked the beneficent earth. Yet even agriculture was not free of risks

[36] M. Rostovtzeff, *The Social and Economic History of the Roman Empire*², (Oxford, 1957), i; Chs. 2 n. 18, 5 n. 16.

[37] J. H. Thiel, *Studies on the History of Roman Sea-Power in Republican Times* (Amsterdam, 1946), 16.

and Cato seems to have found it more profitable to take risks at sea than on the land, where bad weather could ruin crops (*Cato Mai*. 21. 5). The idea of comparative risks, as well as comparative profit may also be relevant to Pliny's comparison between viticulture and sea-trade in *HN* 14. 52, since the vine's profit potential was balanced by its need for care in cultivation.

In view of all this, it seems unlikely that Pliny would want to condemn trade as a motive for seafaring. Yet, in *HN* 2. 118, he reproaches voyages made 'for the sake of profit not knowledge'. The antithesis is not as simple as it looks. In the preceding paragraphs, he had contrasted the thirst for knowledge in adverse circumstances with its neglect in times of peace and encouragement; in § 118 he approves of the selflessness of the motives of earlier men: 'the rewards were not greater then' and 'they made the discoveries with no other reward at all than the achievement of helping posterity'. But his real lament is for the decline in the quality of men's minds; the idea that 'old age has overtaken the characters of men, not their revenues' is related more specifically to mental exercise when he continues the remarks on voyages by saying that 'the mind blindly engrossed with avarice does not reflect that knowledge is a more reliable means even of making profit'. Pliny is in favour of voyaging for commercial gain if, in accomplishing it, man uses his intelligence, applying knowledge already gained and adding to it by personal experience to help himself and others in future voyages. He is against mere greed, the *caeca mens* which mindlessly follows the results of others' groundwork 'now that every sea has been opened up and every coast offers an hospitable landing'.

With his interest in *mirabilia* and in discoveries for their own sake, Pliny would have approved the more disinterested motives of those Greeks, such as Solon or Pythagoras, or later Posidonius, who went on voyages of sightseeing and scientific observation. However, he is very conscious that many discoveries at sea were made as a result of military or mercantile aims. He reflects the predominance of such motives in his narrative. His real concern in *HN* 2. 118 is, however, the way ageing *mores* and *caeca mens* have so narrowed and devalued some men's intellectual vision that they cannot see commercial gain in the wider context of *vita*. Later, Ptolemy suggested that some accounts of voyages kept so narrowly to their mercantile objectives that they were misleading.

Some traders even invented dangers *en route*, such as monsters, in order to protect their commercial monopoly.[38] Pliny would, had he recognized the fiction, have condemned its perpetrators for their blind greed.

C. THE IMPERIAL IDEALISM OF PLINY ON SEAFARING

Even Lucretius could look optimistically at navigation in the context of man's development and achievements. It becomes one of the skills expressing man's triumphant march of progress (5. 1448–50, cf. 333–4). Such an attitude was, however, given a more contemporary significance by Augustus' reform and rebuilding of the navy and general maritime expansion. Even now in the early empire, ideas of impiety and the moral strictures placed on trade that were often linked with these cannot be dismissed as mere commonplaces.[39] Stoic thought could accommodate traditional criticisms of cupidity and luxury and give them new life. Pliny's and Seneca's comments on luxury are genuine. We shall see in the next section how they reconciled them with their more positive pronouncements.

We turn to the expressions of enthusiasm for contemporary Roman progress in seafaring in Pliny and others. A passage of Seneca (*NQ* 5. 18. 4–16) places much emphasis on the fact that man can both use and abuse the sea. The idea is broadly similar to those of Pliny on the use and abuse of *ratio* in connection with seafaring. The manner in which Seneca dwells on it, especially in §§ 14–16, suggests that the extent of man's power over the ocean was a relatively new conception. He seems to be trying deliberately to combat simplistic ideas of 'impiety' or of 'unfriendly elements'. It seems right to see here the influence of increased confidence at sea in the first century.

Within this framework of contemporary interest, however, there are different degrees of appreciation of the dangers involved, and of the way knowledge has to be built up by experience. We have seen that Pliny considers that the real mark of rashness at sea is acting without making judicious use of one's own knowl-

[38] Ptolemy, *Geog.* 1. 4. 2; see E. H. Warmington and M. Cary, *The Ancient Explorers*[2] (London, 1963), 229.

[39] *Contra* E. de Saint-Denis, *Le Rôle de la mer dans la poésie latine* (thesis, Lyon, 1935), 300–5.

edge and that of others. This is related to his beliefs in the supremacy and potential of man's *ratio*. The intellectual and practical qualities of a general seemed to Pliny to have value in agriculture: the similar qualities of a commander of a fleet might equally well be applied to the strategy of seafaring generally. It is worth remembering that Pliny ended his military and official career as Commander of the Fleet at Misenum and a practical interest in the sea may sometimes be discerned in the *HN*.[40] The poet Manilius takes a very positive view, celebrating man's conquest of the sea through his advance in knowledge. It was a gradual process; man's belief that he knew enough even before he tried seafaring (*Astr.* 1. 78) is implicitly rejected when the poet goes on to say that the ages sharpened man's wits and he acquired knowledge of agriculture, 'and the wandering sailor penetrated the uncharted sea and established trade-routes between lands unknown to each other' (*Astr.* 1. 87–8). However, the emphasis is on the successes, not the setbacks. The passage continues the praise of the ascent of man's knowledge even up to the stars. Enthusiasm for the linking of the lands reflects not only Stoic enthusiasm for the intercourse of mankind (see below) but also for the very feat of expanding knowledge, with its implied power over what was formerly distant and unknown. When he writes that '(the man born at the rising of Argo) will crave to sail over the entire ocean with his fleet and to visit foreign climes and the deep Phasis' (*Astr.* 5. 43–5), Manilius implies a spirit of pure maritime adventuring not normally associated with either Greeks or Romans. Seneca is more cautious: in *Medea* 318 ff., he surveys the gradual progress man had to make in coming to grips with the sea and he does not ignore the commonplace idea that, by going to sea at all, man was gratuitously providing a new source of fear for himself.[41]

[40] e.g. 2. 128, on nautical manœuvres. Bk. 16 contains much information on timber in general, and a number of references to its use in shipbuilding, together with other tree-derived products, such as pitch (*HN* 16. 42, 52, 56, 158, 195 ff.). *HN* 16. 5, concerning the hazard posed to the Roman fleet by fallen trees near the territory of the Chauci, may reflect Pliny's own service in lower Germany at the time of Corbulo's campaign against the Chauci in 47 (see R. Syme, 'Pliny the Procurator', 218 ff., after Münzer). For Pliny's command of the Fleet at Misenum, see Pliny, *Ep*. 6. 16. 4.

[41] Costa (n. 32), 103 (*Med.* 338–9), a commonplace from Augustan times onwards. He compares Prop. 3. 7. 31–2. For the only occurrence in Pliny (19.6), see 2. A.

Both authors are to be contrasted with Pliny. He has the confidence of Manilius, but, unlike him and Seneca, looks only at the present state of achievement, rather than at the process by which it was built up. He celebrates a contemporary situation in which all Nature, including the sea, is subservient to man. If there is ambiguity in the relationship between man and Nature, this is not allowed to interfere with the specific practicalities of seafaring. In this age, man has the necessary knowledge, either in himself, or to be acquired from contemporaries or predecessors, to undertake fairly safe voyages. Pliny does not bother to dwell on a time when man's knowledge of the sea was imperfect; when, by attempting a voyage, he brought upon himself another *metus* and was, in fact, acting irrationally in a state of ignorance. Seneca comments on this traditional picture, especially in his tragedies, as in his description of the Golden Age in *Hipp*. 530: 'not yet did the trusting (*credulae*) ships plough the sea'. Yet he too, in the *Medea* passage quoted earlier (318 ff.), acknowledges a definite change between past and present: 'now in our time the deep has given way and submits completely to law' (364–5). The conquest, however ambiguous in its early stages, has been achieved. He ends on an optimistic note, the linking of nations and the future discovery of new lands.

As an exponent of a contemporary interest and confidence in navigation, Pliny allows his enthusiasm to overrule his feeling that land was the only element unambiguously benign towards man. Dionysius of Halicarnassus (1. 3. 3) had already depicted Rome as mistress of every sea not only inside the Pillars of Hercules but also of the Outer Ocean. This is particularly interesting since the Outer Ocean, conceived as a kind of river running around the edge of the world, traditionally aroused the greatest fears of the sea. The Elder Seneca provides evidence of the commonplaces on its nature (*Suas*. 1. 1–16).[42] His quotation of Artemon (§ 11) contrasts Ocean with the ordinary inland seas or rivers: it was earth's boundary, the border of Nature itself, the oldest element, and the birthplace of the gods, all of which made it too holy for sailing. His quotation of Moschus (§ 2) mentions the idea that darkness prevails in this region: 'a horrible darkness

[42] On the traditional ideas lingering into Hellenistic and Roman periods of exploration see J. Ramin, *Mythologie et géographie* (Paris, 1979).

oppresses the waves . . . What Nature has removed from men's sight is shrouded in everlasting night.' There is a fear of monsters in the deep (*immanes beluas*, 4). Ocean could even be viewed as a primeval element, outside creation proper, an unknown waste in which rudimentary life is still being formed: 'all that is primitive and incomplete in nature has retreated to this far refuge' (ibid.). This idea is reflected in the bombastic phrase Seneca quotes at the end of this discussion, in which the Macedonians are said to be entering 'Chaos' not simply 'Ocean' (§ 16).

However, Seneca picks out a fairly recent writer, Albinovanus Pedo (§ 15), as an exception to the generally weak and predictable literary descriptions of Ocean. This poet records a contemporary historical event, the expedition of Germanicus.[43] Although he expresses the usual idea of fear arising from the darkness and the monsters, he is commemorating an actual voyage, not debating as to whether it should happen at all. By the time that Pliny is writing, there are more signs of this increased familiarity. In practice, our author does not distinguish between inner seas and outer Ocean, except where he discusses the size of the *mundus* as a whole (2. 173). In *HN* 27. 2, he echoes the idea that Nature in its fullest sense did not extend beyond the Pillars of Hercules, but the context is the *immensa Romanae pacis maiestas* which allows commerce with even that area. Ocean's monsters become objects of curiosity rather than of fear in the first century, and this is particularly noticeable in Pliny, 9. 9, who records that Tritons and Nereids were reported to Augustus (9. 11) and Tiberius and that 'elephants' and 'rams' were washed up on the Gallic shore (9. 10).[44] The primeval fertility of the sea is also a source of fascination rather than of fear.

[43] W. A. Edward, *Commentary on Seneca; 'Suasoriae'* (Cambridge, 1928), suggests that this was the elder Germanicus' expedition of 12 BC rather than the younger's (AD 16), which was marred by disaster (Tac. *Ann.* 2. 34). On the other hand, Pedo may have been present on the latter expedition (*Ann.* 1. 60), and his poem is sometimes seen as influencing parts of Tacitus' account, e.g. *Ann.* 2. 34. The evidence either way is not conclusive. It need not necessarily follow that the mention of a Pedo in Tacitus proves that the poem was about the second expedition. According to Edward, the first would have made a greater impression on contemporaries. Certainly, Tac. *Germania* 34 seems to regard this expedition as the only real attempt at exploration in the area to date. He makes comparatively little of it, however, although this may perhaps be due to its incidental nature in his description of Germany.

[44] Marine *mirabilia*: see Intro; p. 9. Augustus collected the bones of such monsters at his villa on Capri (Suet. *Aug.* 72).

But in the sea, lying so widely outspread and so yielding and productive of nutriment because the element receives generative causes from above and is always producing offspring, a great many monstrosities are found. The seeds and first principles intertwine and interfold with each other now in one way and now in another, now by the action of the wind and now by the action of the waves. Thus the common opinion is ratified, that everything born in any department of nature exists also in the sea, as well as a number of things not found elsewhere. (*HN* 9. 2.)

There is a mystery and ambiguity surrounding this fertility, however; the sea is prone to excesses ('a great many monstrosities are found') to which the useful, beneficent earth is not.

The distinction between inner and outer seas was not the only one to lose importance. As Roman control of the sea grew, that element might be regarded increasingly as merely the route to more land, rather than a danger. The distinction between the two elements was in one sense being worn away. As the Romans conquered distant lands, so too might they conquer the sea. Significantly, perhaps, it was Claudius, the emperor who attracts the most favourable comments from Pliny, who displayed most interest in trade and navigation generally, and in the idea of the conquest of the sea. Suetonius (*Claud.* 17) tells us that, on his return from Britain, his triumph included a naval crown representing 'the crossing and conquest, so to speak, of the ocean'. A similar interest is indicated by the reference to the 'glory of *imperium* carried beyond Ocean' in the speech of Claudius preserved on the Lyon tablet.[45] Pliny himself concentrates on the naval aspect of Claudius' triumph for the successes in Britain; the emperor sailed into the Adriatic from the north of the Po in a ship which was 'more like a house than a ship' (3. 119). It has been observed that the fact that Pliny lists the known seas in a catalogue reflects the Romans' feeling that they had subjugated the sea.[46] It may be a reflection of this confidence, as well as of his own career in official government, that Pliny sometimes uses Roman conquests in a specific area as a basis for his gazetteer of the world.[47]

[45] E. M. Smallwood, *Documents Illustrating the Principates of Gaius, Claudius and Nero* (Oxford, 1967), no. 369, ll. 39–40.

[46] Saint-Denis (n. 39) 440.

[47] For Pliny's interest in the Romanization of North Africa, see J. Desanges, *Pline l'Ancien, Histoire naturelle, livre V* (Budé, Paris, 1980), 20. For a new view of the date of some of his source-material, see B. D. Shaw, 'The Elder Pliny's African Geography', *Historia*, 30 (1981), 424–71.

Roman imperial control over far-flung lands is facilitated by conquest of the sea which divides them. Combined with this patriotic interest is a personal enthusiasm which can be seen in certain emphases in the *HN*'s composition. Foreign writers on navigation are not considered inferior to Roman informants, even if Pliny does consider Roman military penetration to be the greatest aid to discovery.[48] He accepts the information of Pytheas, of whom Strabo and Polybius were sceptical on the grounds that he was an insignificant Massilian. The evidence of the Carthaginian Hanno is rejected, but only because Pliny could find no trace of the towns he claimed to have founded (5. 8).[49] He does not query the expedition itself and wants to believe that Hanno reached the Gorgades islands, although he admits that 'everything in that area is uncertain' (6. 201). The whole of the gazetteer (bks. 3–5) is an ambitious attempt to write a popular summary of the world, and must reflect personal enthusiasm as well as a general contemporary interest. It aims to give a general reader as much information about the places of the world as is possible without becoming unreadable or boring. This is clear from such phrases as *HN* 3. 28, on the listing of tribes which can be mentioned 'without inducing boredom'; 3. 2, an assertion of brevity and lack of detail other than names; 3. 139, 'few of these peoples are worth mentioning or have pronouncable names';

[48] Military sources as the most reliable; expeditions as the prime means of exploration in Pliny: 5. 51 (the source of the Nile has not been found, partly because 'it has only been explored by unarmed investigations without the wars which have discovered other countries'). Unlike Seneca and Lucretius, Pliny does not moralize on the use of the sea for hostilities. In 6. 23, he says his account of Asia's interior will differ from the older writers' because he has devoted much care to ascertaining thoroughly recent events (*nuper*) from Domitius Corbulo's expedition, from suppliant kings and hostage children. Cf. also Aelius Gallus' expedition to Arabia (6. 160), Augustus' fleet's information on the Northern sea (2.167), and the part played earlier by Alexander's eastern victories in opening up a large part of the world (2. 168). Mistakes made by expeditionary forces are discussed and refuted in detail, the impression being that Pliny has pounced on a small point in otherwise good accounts, e.g. 6. 40, where Corbulo's mistakes are rectified by reference to Alexander's expedition.

[49] On Pytheas, see 2. 187; 217, cf. 1, index to bks 2 and 4; Strabo 2. 4. 2; 2. 3. 5. See J. Ramin, *Le Périple d'Hannon* (BAR, suppl. ser., 3; Oxford, 1976), 65 and n. 17, who appears to class Pliny with Strabo and Polybius on this question of bias but points out that Pliny does not doubt that the expedition took place. Perhaps some of the cities he looked for in vain were trading-posts of an impermanent nature, cf. Ramin's commentary, *passim*, on the difficulties of identifying Hanno's route and his aim.

and again 3. 28, which implies a similar reservation, 'a collection of peoples . . . who are unimportant and have outlandish names . . .'.[50]

Finally, it is important not to overlook the idealistic aspect of Pliny's interest in Rome's growing familiarity with the sea. His view of the Roman empire has a strong element of philosophic idealism in it. He draws parallels between the rule of Rome and the rule of *Natura* in *HN* 27. 2–5 and portrays commerce as a means of general human welfare, not just practical profit-making. *Humana salus* (27. 3) is helped by the commerce in healing plants (see below, next section; and Ch. VI. 2. E). This idea has its roots in the principle of self-preservation discussed by Cicero in *De Finibus* 4. 16–17: 'All Nature desires to preserve itself . . . With this object, the philosophers declare, man has called in the arts also to assist Nature and chief among these is the art of living, which helps him guard the gifts which nature has bestowed and obtain those that are lacking'.

We see this principle in Pliny's descriptions both of *artes* and of natural features. He says of Italy that she is open to *commercium* and the land itself actually rushes into the sea as it were to help mankind (. . . *tamquam iuvandos ad mortales ipsa avide in maria procurrens*, 3. 41, cf. sect. 2 D). Of the art of navigation, he remarks: 'by what greater skill of men has nature been aided in any part of herself than by sails and oars?' (32. 1). Such remarks are, moreover, integral to his own personal philosophy of man and nature. Teleology and Plinian optimism ensure that the impiety idea is obscured by emphasis on the positive aspect, co-operation of man and *Natura*. Flax, attacked in *HN* 19. 3 for providing sailcloth for rash navigators, becomes one of the basic commodities 'which life should not lack' (37. 202). The patriotic

[50] Much has been written on the correctness or otherwise of Pliny's geographical details. The consensus seems to be (e.g. J. O. Thomson, *History of Ancient Geography* (London, 1948) 226–8, 323–5 that Pliny is confused and obscure, a defect which was not helped by his collation of different sources (O. A. W. Dilke, *Greek and Roman Maps* (London, 1985), 68). Too much detailed criticism of this kind, however, is ultimately of limited usefulness. More importantly, it seems to miss the main point: Pliny is providing a general picture of the world as a whole and for this reason restricts himself to giving 'bare names' of places (3. 2), cf. Dionysus Peregetes 1166–8, who restricts himself to major places to give a view of the world as a whole. See also K. G. Sallmann, *Die Geographie des älteren Plinius in ihrem Verhältnis zu Varro: Versuch einer Quellenanalyse* (Berlin and New York, 1971), 191.

and philosophical ideas are often identified. As well as the parallel between Rome and *Natura* as rulers, Italy is called 'the ruler and second mother of the world' (37.201). The attribute which has made her so is her harmonious combination of man and his *artes* with her natural features and commodities, which is illustrative of this overall co-operation of man and Nature.

D. *LUXURIA* AND *COMMERCIUM*

Stoic thought was undoubtedly opposed to luxury. However, it did not follow that all dealings between different parts of the world, whether by land or sea, were to be condemned. *Commercium*, in its wider sense of intercourse, was a Stoic ideal. 'We are not enclosed by the walls of one city, but have made contact (*commercium*) with all parts of the globe and proclaim ourselves citizens of the world' (Sen. *De Tranq. An.* 4.4). Yet it was not easy to make distinctions; trade was a prime means of promoting more general communications. *Ius commercii* (Ulpian, *Reg.* 19.5) was the technical term for the law of acquisition not only between individuals but also between different nations. When Seneca praises the benefit of rivers which offer a route *ad commercia* (*De Ben.* 4.5.3, see below, sect. 2. E), this must refer primarily to trade. In *NQ* 5.18.4–16, the *commercium* that is condemned is the aggressive intercourse of war between nations. As we saw (sect. 2. C, *init.*) Seneca here makes the obvious point that it is the purpose to which 'commerce' is put which determines whether it is good or evil. This would, no doubt, enable him to condemn certain types of trade, such as that in luxuries.

We have seen that Pliny's ideal of human *commercium* generally is inseparable from its more specific meaning of trade. Commerce by sea arouses a greater enthusiasm in him than it does in Seneca. When we come to consider his attitude specifically to the luxury trade, it becomes clear that his undoubted dislike of *luxuria* is, to a certain extent, tempered by a more open pragmatism. Luxury in itself is not the greatest evil. It is certainly true that moral condemnation of *luxuria* is more than a commonplace to Pliny. Under Tiberius, it had been a matter of serious debate (Tacitus, *Ann.* 3. 52–4). Tacitus thought that the personal example of the emperor had most success in combating excess, as his opinion of Vespasian's character suggests (*Ann.* 3.55.5). Vespasian's parsimony could be seen as characterizing his age.

There is, perhaps, an implicit contrast with the previous principate in Pliny, for a large number of his attacks on luxury are in the context of Nero's reign (see Intro. 17–18). Like Tacitus, he also noted the contrast of the present with the late Republican era in this respect (36. 8). Vespasian's policy was by no means based solely on idealistic grounds; it was a matter of practical necessity since the treasury had been emptied by Nero's extravagance and by the civil wars of 68. Moreover, Vespasian displayed a natural talent—some said a positive zeal—for the task of rebuilding Rome's finances. His astuteness was notorious (Suet. *Vesp.* 16, 19, cf. 4). As the East became more accessible, its products were easier to acquire and Vespasian exploited the situation. He was adept at manipulating trade to the advantage of himself and the state treasury. On the one hand, it may well be that the establishment of the Horrea Piperatoria was his rather than Domitian's measure;[51] on the other, it has been suggested that the comparative scarcity of coins of Vespasian found in India means that he also had some success in checking the outflow of coinage from Rome.[52]

Vespasian's combination of plain living with financial sharpness better explains why Pliny has an anti-luxury stance but an interest in trade which goes beyond the acquisition of useful basic commodities like grain (but see below, on 34. 6). When he expresses moral horror at the money spent on luxuries from India, Arabia, and China (12. 83–4), his censorious generalization is perhaps influenced by his specific example, Nero. In contrast, an earlier reference (6. 101) to the money going to India, in a description of the trade-route between Egypt and India, contains no moral criticism.[53] He need not have given so many details of the route if

[51] T. Frank, *An Economic Survey of Ancient Rome*, v: *Rome and Italy of the Empire* (Baltimore and Oxford, 1940), 231, sees Pliny's long lists of spices as evidence of some actual state fiscal control. Cf. Suet. *Vesp.* 16. 1, the fiscus bought and sold for profit.

[52] Frank (n. 51), 232. M. P. Charlesworth, *Trade Routes and Commerce of the Roman Empire* (Cambridge, 1926) 70, sees it as simply a sign of the falling-off of trade. Against this, L. Casson, *The Ancient Mariners* (London, 1959), 231, attributes the decrease in coin-finds to the natives' preference for coins predating debasement under Nero.

[53] On this passage see M. Crawford, 'Economia imperiale e commercio estero', in *Tecnologia*, 207–17, who contrasts its amoral tone with that of *Ann.* 3. 52–5 and Dio Chrys. 79. C. argues that Pliny and Dio speak as lone individuals and not for the Roman government when they express an awareness of the Roman empire as an economic unity. Once again, it would seem, Pliny is displaying more individuality of thought than he is normally credited with.

such criticism had been his aim. Rather, one suspects that he was catering not only for his readers' curiosity but also for their financial interests. In *HN* 2.118, he had cited 'the thousands of people who go on voyages for profit rather than knowledge' as a reason for giving such a detailed account of the winds. What the verb *exhaurio*—'drain off'—in *HN* 6.101 does denote is anxiety on economic grounds. The draining of coin would apply to Italy in particular, where exports were not balancing the deficit.

It is true that Pliny writes of the immorality of increasing the appetite by means of pepper or of clothing the body in silk (12.29; cf. 6.54), voicing what Tacitus sees as the genuine attitude of those from the provincial towns of Italy (*Ann.* 3.55). Yet, overall, he refrains from wholesale condemnation of the luxury trade, as he does of luxury in general (Ch. II.4). Here again, pragmatism in the form of financial awareness colours his moral criticisms. The inflated prices of luxury goods gave the buyer a false sense of their material (and, in the case of medicines, physical—see below), as well as of their moral value. Indian luxuries, unnecessary frivolities, were sold at 100 times their initial cost (6.101).

The ancient system of bartering, which did without gold, the source of money, in trade, is praised by Pliny in *HN* 33.6. In 6.88, he praises the Sino-Sinhalese system of bartering. 'Commodities were deposited on the opposite bank of the river beside the goods offered for sale by the natives and they took them away if satisfied by the barter. Hatred of luxury is in no circumstances more justifiable than if the imagination travels to that area and reflects what is procured, at what price (*quo*), and for what purpose (*quare*)'. Pliny invites a contrast between the nature, price, and purpose of these bartered goods and Roman luxury. The implication is that the former are, perhaps, more utilitarian than luxurious. But above all, it suggests the exchange is fair and level-headed; goods are exchanged for goods, not for vastly inflated sums of money. There is none of the irresponsible playing with money criticized by Pliny in his passage on the refinements of purple dye (*HN* 9.139 ff.). Indeed, the false ingenuity described in that passage, the preoccupation with refining what was already luxurious (see Ch. II.4), is also absent from Sinhalese culture. The Sinhalese possess luxury but are not irrationally obsessed by it. 'Not even Ceylon lacks the vices which belong to us. Gold and silver are valued there also, and a kind of marble which resembles tortoise shell and pearls and precious stones are held in honour. In fact, the whole mass of luxury is there carried to a far greater pitch than ours. They said there was greater wealth in their country than in ours but that we made more use of our riches' (6. 89).

The false values Pliny sees as resulting from high prices are illustrated by his comments on luxurious medicines. He makes a point about human psychology in *HN* 29. 28, when he says there are people who 'do not believe things are helpful unless they are also expensive' (see Ch. VI. 1. B). In *HN* 22. 118, he criticizes mysterious and ingeniously mixed medicines from Eastern products as being luxurious and medically useless (Chs. II. 2, *fin*; VI. 1. B). Such ingredients may, if wished, be bought to make *deliciae*. Goods should be purchased in the full recognition of what they are in terms of the moral and material essentials of life. The cheapest and simplest things are often the most beneficial, a point Pliny never tires of making in connection with herbal remedies. Thus, it is at the international trade in plants (27. 2–3) that the greatest praise of commerce in Pliny is aimed. Yet the less admirable trade in luxuries is still allowable, provided that the true sense of values, as outlined, is preserved.

Pliny believes that, to a certain degree, the reasoned effort and planning applied to farming can now be applied to contemporary seafaring. The seas are now being opened and tamed. However, there is a regularity in the conscientious effort implied by *sedulum* in 14. 52 (above, Ch. V, intro.) and there is a rationality which must rule *labor* generally in farming. These cannot be wholly reconciled with a luxury trade involving irrationally inflated prices and the greater capacity of the sea as opposed to the land for unreliability. The risk involved in obtaining goods by or from the sea itself increases their prices, making 'luxury' even more 'luxurious'. Pliny makes this point in *HN* 22. 3–4, when he contrasts the *murex* dye with that of plants gained more safely and cheaply on land. He declares his intention of drawing attention to these humble plants, and thus curbing luxury by cheapening it.

In contrast to the rational work-ethic in farming, *desidia* is often paired with *avaritia* (14. 2–5), and *avaritia* with *luxuria*, since so often the wealthy spend their money on indulgence in non-essentials. Idleness, greed, and luxury form a vicious circle. *Desidia* can involve allowing others to act for the individual and usurp the functions of his mind (29. 19). Pliny criticizes greedy doctors for doing this in *HN* 29. 24–7. They use luxurious treatments to increase their hold on their patients: '. . . none of life's seductive attractions have they overlooked. I am inclined to believe that our ancestors were displeased with imports from abroad and the fixing of prices by foreigners' (29. 24, cf. Ch. VI. 1. A–B). The whole process is a negation of *vita*: 'pleasure begins to live and life itself has ceased to exist' (14. 6). If it is not to have a debilitating effect on both mind and body, luxury must be properly evaluated. *Ratio* must be applied not only to the method of acquiring it, but also to ensure recognition of its real value and effect.

E. *AQUA AMOENA ET NAVIGABILIS*: LANDSCAPE AND NAVIGATION

Not only the sea was used for navigation. It was an important function of rivers too. Rivers, *flumina*, being less *incerta*, finite bodies of water surrounded by friendly *terra*, illustrate how Pliny's preoccupation with man's activity colours his description of the natural environment in which it takes place. The two words most often used by Pliny to describe rivers are *amoenus* and *navigabilis*. Both signify affinity to man. The former, a quiet pleasantness[54] rather than the wild beauty of mountains or sea, denoted the kind of scenery which most appealed to Greeks and Romans (see below). Inland waters were a regular feature of traditional descriptions of the *locus amoenus*[55] in poetry and

[54] See *TLL*. s.v. *amoenus*. Associations with peace and quiet: Cic. *Att.* 4. 8. 1, 'nowhere is quieter, cooler, more pleasant (*amoenius*), sc. than Antium'. Such a spot may be secluded: Cic. *Att.* 7. 20. 2, '. . . a remote, healthful, pleasant place . . .', but not in the sense of wild or deserted. There is a tameness and affinity with human civilization: see Cic. *S. Rosc.* 133, on countryside which is *amoenum et suburbanum*. Pliny himself speaks of '. . . the lake of Genesareth, skirted by pleasant towns' (5. 71). The 'tamed' atmosphere of *amoenus*, together with its frequent association with water and shady trees, was epitomized in the garden, rather than in really wild nature (Ch. II. 5 and nn.).

[55] Descriptions of rivers and fountains as *amoenus* in Latin literature are very numerous, e.g. Lucr. 4. 1024, Livy 45. 27. 10, Virgil, *Aen.* 7. 30, Hor. *Od.* 3. 4. 7, *Ep.* 1. 106, Ovid, *Met.* 14. 331, Apul. *Met.* 4.17. The tradition continues into the 4th c. in Ausonius' celebration of the Moselle (*Mos.* 21). Enthusiasm for the natural beauty of a river is also to be seen in Tiberianus' short poem *Amnis ibat*.

standard rhetorical exercises. Pliny mentions one of the best known, the Peneus in the vale of Tempe, in *HN* 4.31.[56] Elsewhere in the *HN* he will often mention the 'pleasantness' of a river or lake (3.20 *amoeno stagno*; 5.88 *aquis amoenis*). A more utilitarian tradition of description[57] is to be seen, for example, in Xenophon's *Anabasis*, book 1, where the width of the river and its relationship to towns and fortresses is generally given: such detail is relevant to the military context. While they may have derived partly from the interests of his geographical or navigational sources, Pliny's frequent descriptions of rivers as *navigabilis* also reflect his personal interest in commerce.[58] In general, rivers and lakes were seen as suitable material for use in speeches praising a city; Menander Rhetor (349.25 ff.) says that, like other features, they should be praised with reference to their pleasantness and usefulness.

More interesting is the way in which, for Pliny, the utilitarian aspect combines with aesthetic appreciation: the latter is increased by the former, as his admiration of man-made splendour in the form of public buildings also suggests: the Forum of Augustus and Vespasian's temple of Peace belong among 'the most beautiful (*pulcherrima*) buildings the world has ever seen' (36.102). In contrast, large ornate private houses (36.5–6, 110–12) and theatres (36.113–20) incur Pliny's moral censure.[59]

[56] See C. J. Fordyce, *Catullus: A Commentary* (Oxford, 1961), 312.

[57] Details of width are given e.g. 1.4 (Psarus, Pyramus, Dardes); 2.4 (Zapatas); 4.3 (Centrites); 6.2 (Lycus). Width and name of city on or near the river, e.g. 1.2 (Maeander and Marsyas; Cydnus); 1.4 (Euphrates, Carsus); 1.5 (Mascas); 2.4 (Physcus). The general importance of the practical use of rivers is reflected e.g. In Sen. *Ad Helv*. 9.1 '. . . no large or navigable rivers . . .'.

[58] *Navigabilis* and *amoenus*. The Romans derived *amoenus* either from *amor*, love, or from *amunus*, 'giving nothing'/'unproductive'. Both traditions are described by Servius (*Aen*. 5.734 and 6.638). In one sense, therefore, *navigabilis*, with its connotations of trade, is the antithesis of *amoenus*, productive of nothing but pleasure, a contrast suggested by Sen. *Ben* 4.5.3. 'These rivers surround the fields with their pleasant (*amoenissimis*) curves; those, with their wide and navigable (*navigabili*) course offer a route to trade.' It is noticeable that Pliny does not use the specific words *amoenus* and *navigabilis* of the same river. However, it was more usual to associate *amoenus* with the general productivity and fertility of a Nature who was friendly, and so pleasant, to man (see *TLL* for the connections of such words as *fructuosus*, *fecundus*, *dives* with *amoenus*). In fact, it is clear that Pliny identifies rather than contrasts the aesthetic with the utilitarian aspect of rivers (see following paras.).

[59] Buildings for pleasure are also contrasted with purely utilitarian wonders of engineering, 'marvels unsurpassed by virtue of their *genuine* value' (121);

A river that is *navigabilis* will also be of sufficient size to be imposing: man can master it but also admire it. Pliny writes appreciatively of the *clari et navigabiles* tributaries which flow into the Danube (3. 147) and of the River Sebou in Mauretania, *magnificus et navigabilis* (5. 5).

The description of the River Jordan identifies the idea of pleasantness and of co-operation with man. 'It is a delightful (*amoenus*) stream . . . wandering about . . . and putting itself at the service of the people who dwell on its banks . . .' (5. 71). The River Tiber is a particularly good example of this harmonious relationship of rivers with man's activities and his wider adaptation of the natural landscape: '. . . it is a most tranquil trafficker in the produce of the earth with perhaps more villas on its banks and overlooking it than all the other rivers in the whole world. And no other river is more circumscribed and shut in on either side; yet of itself it offers no resistance, though it is subject to frequent sudden floods, the inundations being nowhere greater than in the city itself. But in truth it is looked upon rather as a prophet of warning, its rises being always construed rather as a call to religion than as a threat' (*religiosus verius quam saevius*, 3. 54–5). Despite its restriction by the works of man, Pliny insists that the Tiber's floods are not a sign of resentment. When a river floods, it can even be seen as an ally of *Terra*, bestowing extra fertility on the soil. Thus, while the River Po damages fields, it does not carry off the soil, but deposits it in the form of a richer silt; 'it claims no part of its plunder for itself and where it deposits its spoil it bestows fertility' (3. 117–18).

Even more striking is Pliny's occasional transference of this harmony between rivers and man's activity in nature to the sea. We have seen that his interest in *commercium* led him to depict Italy's land as running into the sea to welcome it (3. 41, see above, sect. 2. c). In the same passage, Italy's wealth of inland waters is equally balanced by her maritime advantages: '. . . the rich supply of rivers and fountains all over its surface, its many seas and harbours, and the bosom of its lands offering on all sides a welcome to commerce . . .'. This may be compared with the maritime emphasis in his *laudes Italiae* in *HN* 37. 201: 'Italy

aqueducts, the draining of the Fucine lake, and the building of the harbour at Ostia (121–5). This is also relevant to the utilitarianism Pliny displays in his sense of values and in his approval of man's action in the natural landscape, cf. n. 61.

excels in the easy access she offers to all peoples, her shores with their many harbours, and the kindly winds that blow upon her'. This is an emphasis notably absent from the passage of Virgil (*G.* 2. 136 ff.) which inspired it. Virgil reserves his detail for the inland features—rivers, lakes, and the man-made sea-barriers—barely mentioning Italy's seas. Lake Benacus is itself compared to a miniature ocean, as though Italy contained even a safe version of the sea within her boundary. The sea proper is depicted as a would-be invader of the land, held back by the barriers. There is no interest in commerce, and perhaps even implicit criticism of the Eastern luxury-trade: the remoteness of the East is emphasized (171–2) and the unwarlike inhabitants are to be repulsed rather than associated with.[60] Pliny's picture, in contrast, stresses that Italy is an active ruler of Empire, literally reaching out for mutual commerce with her scattered subjects. Her sphere of influence is world-wide (27. 3).

Such a description not only ignores the difficulties normally associated with seafaring: it also contradicts the physical actualities. The Mediterranean winds were not always kind. Except for the Greek settlements in the south, Italy was not particularly rich in harbours, as the attempts to build an artificial one at Ostia suggest. The harbour, however, being a shoreline feature where *Terra* meets *Mare*, marks the point at which man can exercise some control[61] over the sea. The man-made harbour is analogous to the man-made inland waterway, which exercises greater human control over the path of a river (3. 119). The security of

[60] Virgil's ideology: see n. 3.

[61] Cf. *Ann.* 1. 79 for evidence of the attitudes towards the engineering of nature by man with regard to the diversion of the Tiber tributaries. Tacitus lists superstition, as well as technical difficulty and the complaints of towns endangered by flooding if the diversion took place, as possible reasons for the senate's decision to drop the plan. Certainly, a *contra naturam* argument was used by the Reatines in their objection, but only, it seems, to bolster their main concern, safety. Pliny only once suggests that a feat of this type is *contra naturam* and that is with regard to the canal on the Isthmus of Corinth (*nefastum*, 4. 10). As it is the sole example, it may well be a rhetorical use of what had become a commonplace concerning this particular scheme, doomed as it was to constant failure (4. 10–11 and Daremberg–Saglio II. ii s.v. fossa (Paris, 1896), 1330), most recently under Nero. M. Frederiksen (n. 11) cites this passage of Tacitus as evidence that he had an aristocrat's disdain of technology, in contrast to Pliny's equestrian interest, cf. their contrasting treatments of the Fucine lake scheme (*Ann* 12. 56–7; *HN* 36. 124). But Tacitus' narrative purpose may equally well explain his 'lack of interest'.

the harbour was a rhetorical commonplace[62] as well as being a basic item of practical information for ancient navigators. Pliny is no doubt influenced by both interests but also follows his personal preferences when the scanty detail he allows in his itinerary of cities and islands is often a reference to their use as harbours. In *HN* 4. 73, Samothrace 'gives the worst anchorage of all' the islands in the North Aegean; in *HN* 5. 133, Syme 'benevolently supplies eight harbours'.

Pliny's perception of landscape and seascape is affected by his interest in commerce. In the sea passages just discussed, with their emphasis on harbours (3. 51, 37. 202, 4. 73, 5. 133), he is also at his closest to what one critic has called the 'sea from the beach' attitude,[63] common in ancient poetry and embedded in aesthetic judgement. If the sea in a wild and unruly state was enjoyed at all, it had to be experienced indirectly, from the safety of dry land or even from a window. For Pliny's nephew, a dining-room whose window is splashed with spray is implicitly a thrill (*Ep*. 2. 17. 5), unlike another room which 'looks on the sea from a greater distance but with greater safety'. The more usual aesthetic preference is stated clearly by Seneca. 'Let a ruler . . . ask himself which condition of the world is more pleasing (*gratiam*) to the eye and more lovely (*pulchrior*)—when the day is calm and clear (*sereno . . . puro*), or when all nature quakes with clash upon clash of thunder and hither and yonder the lightnings flash?' (*De Clem*. 1. 7. 2).

However, the taming of the sea in Pliny is something of an exception to his general view. His is more often the eye of the dispassionate observer of natural phenomena. His work has more in common with scientific geographical writings, where there might be emphasis on the elements and their interplay, a tradition derived from Aristotle's *Meteorologica*. He quotes from followers of the tradition such as Pytheas and Posidonius (e.g.

[62] The harbour, too, was an important feature in rhetorical pieces praising a city. Menander Rhetor (351. 20 ff.) says that praise of harbours should include mention that they are free from waves and wind and are sheltered. If it has many harbours, a city 'extends many hands in friendly welcome to those arriving from the sea', cf. Sen. *Ad Marc*. on harbours man-made and natural: the latter '*Natura* placed . . . as a protection of fleets' (17. 4). 18. 5 describes 'shores curving to form a harbour . . . islands which break up the sea's expanse . . .'.

[63] Saint-Denis (n. 39) 97 ff.; 479, cf. L. Casson, *Travel in the Ancient World* (London, 1974), 231–2.

2.187). His factual discussion of winds in *HN* 2.112ff. may be contrasted with his generalization about Italy's in 37.201 and with the unscientific literary clichés which abounded in poetic descriptions of storms at sea.[64] We can never overlook Pliny's image of the sea as an element, as part of the *dimicatio naturae*. One passage is of particular interest in this respect. It shows Pliny going to the opposite extreme from that explored earlier in the section, and focusing on a Nature in which not only the sea, but other elements too are wild and remote from the civilizing influence of man. Far from softening the sea's image and suggesting its harmony with *Terra*, he can depict even rivers and certain areas of land as being involved in *dimicatio* with cultivated *Terra*.

In *HN* 2.174 (above, Ch. V, intro.), Pliny had included rugged and uninhabitable areas of land and fresh as well as salt waters among the elements invading *Terra*. In *HN* 5.84ff., a river, the Euphrates, becomes an untamed natural force fighting another, the rugged Taurus mountain range. It is the latter which attacks. 'Occurrit ei [the Euphrates] Taurus mons'. At first, the river is stronger, but then, in Cappadocia, 'for the first time in this combat, the Taurus carries the river out of its course and though conquered and cleft in two it gains the victory in another manner by breaking its path and forcing it south. Thus this *dimicatio* becomes a drawn battle.' It becomes clearer later in the passage that the mountain is the wilder of the opponents. The Taurus range now opposes the sea itself in a struggle for mastery over *Terra*: 'in dominating the land it encounters the opposition of the seas . . .' (5.97). Later, the mountain range tries to encroach on ocean: 'it occasionally throws out arms as if trying to invade the sea' (5.98). The passes known as the Armenian, Caspian, and Cilician Gates are apparent concessions to man's needs, but these mountains remain basically untamed: 'even where it gapes open and makes a passage for mankind, nevertheless it claims for

[64] Saint-Denis (n. 39) 212ff., 225, 374–81, 449, 466. It is interesting that Saint-Denis credits Ovid with not making the mistake of most other Latin poets and mixing contrary winds in his descriptions (377–80). If Ovid's descriptions contain little interest in the sea as an element (374ff.), yet he has authentic details of nautical manœuvres (Saint-Denis 377; *Tristia* 1.2.27–30). It may be added that Ovid has a sense of the contest between man's *ars* and the force of the sea e.g. *Met.* 11.494, 'so great as is the impending weight of (the sea's) destruction, so much more mighty than all his skill'; 'Skill fails . . .' (537).

itself an unbroken unity by calling these gates' (5.99). Overall, however, *dimicatio naturae* results in a balance (cf. 'drawn battle', 5.85). Elsewhere, Pliny says that water may dominate the other elements and its activities cause discord, 'making the world wage civil war with itself' (31.2); yet its presence is vital to the coherence of the earth. There is an interdependence of the elements designed by *Natura artifex* (2.166, cf. 2.10–12).

Conclusion

A naturalistic approach to the sea and the other elements coexists in the *HN* with Pliny's generally anthropocentric view of the world's resources. Dangers and ambiguities are recognized in the case of the sea, but the emphasis upon them varies according to the context. The practical and patriotic interest in Rome and her empire predisposes Pliny to look at the advantages of navigation. Patriotism is integrated into the philosophy of man and nature, in which negative aspects, including the impiety traditionally associated with navigation, are minimized. Fascination with man and his achievements is the decisive factor, whether Pliny is dealing with agriculture or with navigation. The land–sea contrast may be seen more as a comparison of the combination of risk and profit involved in agriculture as well as seafaring. This is suggested above all in the passage which introduced this chapter, *HN* 14.52. *Terra*, fertile and beneficent as she generally is, nevertheless offers a continual challenge to her cultivators to make the most satisfactory use of her resources. Bad weather can still threaten the farmer with an insecurity which cannot always be provided against by *ratio*. The context of the passage is that of crops which are exceptional in their yield. Perhaps Pliny feels a certain reticence, even with regard to such agricultural *praemia* (14.47),[65] as he does to a much greater extent with regard to the irrationally high *praemia* of the sea-borne luxury trade. In the case of rivers, the comparative reliability of the surrounding land dominates the instability of the water. In terms of risk and friendliness to man, the river is closer to *Terra* than to *Mare*.

[65] This would be especially true of 14.48–52, since it is in the more specific context of the recent commercial development of viticulture, to which the idea of the 'indecent profit' is particularly relevant. See n. 35 and N. Purcell, cit. ibid.

Mare, though less secure than *Terra*, is none the less not to be feared. It offers a greater challenge than agriculture to man's *ratio* in the weighing-up of considerations of risk and profit: yet, overall, Pliny believes that man's progress has led to a state of contemporary knowledge which enables that challenge to be successfully undertaken. At times, the whole of *Natura*, not just *Terra*, seems to be little more than a servant to man's initiative.

VI

Ars Medicinae: Man's Use of Nature in Medicine

Introduction

The centrality of man in Pliny's universe suggests that Nature's most important task is to support the existence of the human race. Much praise is lavished on her provision of *remedia* which ensure *humana salus*.[1] Here, it would seem, is a promising field for man's activity in nature. To be skilled in the art of healing fellow men should earn Pliny's highest praise. After all, for man to aid man is, by his own definition, godlike (2. 18). Instead, *medici* are the victims of a lengthy tirade, initiated by quotation from Cato and embellished by Pliny himself (29. 13 ff.). However, this passage makes two things clear: Pliny says that it is the *ars*, not *medicina* itself, which is under attack. By this, he means not so much the body of theory, as the body of people who practise it, the medical profession, who consisted mostly of Greeks. Pliny's criticism of medicine is, in fact, criticism of the way man's activity can misuse Nature's most basic gifts, a theme by now familiar from the preceding studies of Pliny's thought. Indeed, by examining how the *ars* can be said to have corrupted *medicina*, we encounter many of the Plinian ideas on the use and abuse of *ratio*, such as the questions of false ingenuity, *caeca subtilitas*, and the balance of theory and practice. The motivation both of the practitioners and of the patients themselves brings up familiar ideas on the right attitude to life generally. These include the contrast between idle luxury and the independent self-sufficiency of the individual, and Pliny's humanitarian ideal.

[1] *Humana salus*; 27. 3; cf. 22. 1; 23. 1; 24. 1 and 4; 26. 10. See Ch. II; intro. and n. 3.

However, another factor also comes into prominence: Pliny's anti-Greek prejudice, a concept familiar from elsewhere in Latin literature,[2] especially in satire. The Roman ethos, which, as we have seen, was also bound up with much of Pliny's general philosophy of life, was felt to be threatened by these foreigners. Their cleverness was seen as perverted ingenuity and their moral outlook was suspect.[3] As the practitioners of *ars medicinae*, they were particularly feared.[4] It gave them power, both that arising from their status as intellectual experts and that arising from their control over the lives of others. It highlighted their arrogance, for example over the use of the Greek language, which made the *ars* more exclusively theirs (29. 17) and must have increased the feeling of powerlessness in many of their Roman patients, even if others were glad to hand over responsibility to these experts (see below, sect. 1. A). Their acquisitiveness had greater scope and their introduction of luxurious and effeminate 'treatments' eroded the ability of their Roman conquerors to rule even themselves as a good Roman and especially a Roman Stoic should (29. 19).

This prejudice brings to Pliny's view of medicine a certain distortion of his ideas. We find that he tries to maintain a divide between a corrupt, Greek, *ars medicinae*, and a Roman herbal medicine that is grounded in his ideal of life in accordance with Nature: the remedies which Nature provides are used in their simplest form by those who really need them. There are no sinister or mercenary motives, only a desire to enhance *vita*. But in emphasizing the corrupt ingenuity of the Greek *ars*, Pliny has to try to ignore the different corruption of *ratio* found in herbal medicine: its frequent lack of any rationality at all, and its

[2] See Intro., pp. 18–20; also J. P. V. D. Balsdon, *Romans and Aliens* (London, 1979); G. W. Bowersock, *Augustus and the Greek World* (Oxford, 1965), Chs. 5 and 10; A. Momigliano, *Alien Wisdom: The Limits of Hellenisation* (Cambridge, 1975), Chs. 1–2; N. K. Petrochilis, *Roman Attitudes to the Greeks* (Athens, 1974); A. N. Sherwin-White, *Racial Prejudice in Imperial Rome* (Cambridge, 1967); A. Wardman, *Rome's Debt to Greece* (London, 1976).

[3] See Balsdon (n. 2), 36–7 and refs. For Pliny's remarks on Cato, see Plutarch, *Cat. Mai.* 23. 3.

[4] See F. Kudlien, 'Medical Education in Classical Antiquity', in C. D. O'Malley (ed.), *The History of Medical Education*, (Berkeley, Calif., 1970), 18 ff., though the question of the slave doctor in ancient Greece is much debated. Despite Domitian's edict, *servi medici* are likely to have continued in many Roman households (Kudlien 19).

association with magic (see Ch. III. 2. A. iv). At the same time, he ignores the common basis shared by the traditional native herbal lore and the more scientific Greek developments. This problem will be examined in the latter part of the chapter.

The disingenuousness of Pliny's professed attitude to 'Greek medicine' becomes apparent as soon as the indexes which make up the first book of the *HN* are opened. Of the vast number of medical sources quoted for books 20–37, Greeks outnumber Romans by at least three to one. Some are, of course, quoted in the text with contempt; for instance the medico-magical works foisted upon the fifth-century-BC philosopher Democritus. But Pliny usually refrains from quoting sources in the actual text. Thus it may be that he found in those same sources unexceptional material which he used without acknowledgement: the Greeks named disapprovingly in *HN* 28. 70 ff. for describing remedies from human bodies and body-fluids may also have provided him with animal or herbal remedies which he accepts without a qualm.

Even more significant is the naming in the indexes of eminent Hellenistic physicians, founders of the medical sects which Pliny scorns (see below, sect 2. B). Erasistratus, who founded a distinguished school at Alexandria in the third century BC, is derided by Pliny for 'a violent change' in theory and for making large profits (29. 5) out of medicine. Yet he appears in the indexes for books 20–7. Moreover, the earlier of these books are concerned almost exclusively with 'herbal' remedies.

Finally, Pliny cannot disguise the fact that there are more Greek than Roman works on herbal medicine (25. 5–8). In spite of the disadvantages he mentions here (see below, sect. 2. D), the illustrated herbals of Crateuas, Dionysius, and Metrodorus must have been significant sources for him. On plants generally, Pliny was greatly indebted to the distinguished botanical writings of Aristotle's pupil Theophrastus.

Pliny's prejudice gives support to the popular caricature of the clever unscrupulous Greek, out to swindle and debilitate both physically and mentally his Roman rulers, perhaps eventually to kill them (29. 14). Yet, if we make allowances for this distortion, his ideas on the preservation of *vita* itself make a fitting conclusion to his study of man's life in nature.

1. *Life according to Nature: Problems with the* ars medicinae

A. INDEPENDENCE OR PASSIVITY? MASTER ENSLAVED TO SERVANT

The position of authority and power attained by the foreign practitioners of medicine at Rome was an affront to Pliny both as a Roman and as a believer in the importance of *Natura*. If *Terra*, in providing *remedia*, and other benefits for *humana salus*, could be called man's servant (2. 155, 157), there was even more reason to keep human doctors in a position of subordination to their patients. Originally, this had been the case at Rome. In the traditional Roman *medicina*, doctors were generally slaves (*medici domestici et familiares*; *servi medici*) or freedmen. In addition, the status of Greek doctors arriving at Rome would not have been very high. Originally, some would have been resident aliens, *peregrini*, some even prisoners of war and hence slaves. Certainly, not all would have had the privileges granted to Archagathus, the first Greek physician to come to Rome and be paid a state wage (29. 12). Cicero, when discussing occupations fit for freeborn Romans (*De Off.* 1. 42), despite recognizing the intelligence required and the particular usefulness of the skill, said that it was fit only for those to whose *ordo* it was suited. However, for the profession to be suited to *ingenui* at all suggests that some doctors at least were now not only freeborn, but even of a respectable middle-class origin.

In any case, their intelligence probably gave doctors an unofficial mobility within the social system from quite early on. Hellenistic medicine had sustained strong links with philosophy; indeed, philosophy had largely determined its development.[5] Thus a practitioner of Hellenistic medicine at its most learned might well acquire friendship among the educated classes and be accepted on a more equal footing than a mere servant would be. Certainly, this happened later, as the doctor friends of Plutarch show.[6] It was learned accomplishments in fields other than

[5] See L. Edelstein, 'The Relation of Ancient Philosophy to Medicine', in *Selected Papers of L. Edelstein*, ed. O. Temkin (Baltimore, 1967), 349–66.

[6] Plutarch's doctor friends: see *Symp.* 3. 4 and *RE* xxi. 1 (1952), 671 and 679 (Avitus and Nicias).

medicine alone which gained the doctor Crito an influential position as Trajan's historian.[7]

Official measures were not, however, long in coming: Caesar's granting of the citizenship to doctors in Rome must have given them greater confidence, both as individuals and as a profession. Under Augustus, doctors moved in the highest court circles, a trend which continued under Claudius and Nero (see below, this section). Domitian's postscript to Vespasian's edict[8] granting privileges to doctors banned the teaching of the art to slaves for money. Eventually, this would have lessened the number of slave doctors. It may have been partly a recognition of a general rise in the social standing of doctors and their wish to safeguard their privileges as those of free men.

Above all, the very nature of the doctor's profession suggests that he could, more than most, achieve an influential position with those who were indebted to him for his services. It is interesting to speculate on the possibility of physicians giving clients in positions of the greatest power and responsibility a certain psychological support and security. Care for their personal well-being would help ease the strain of power. The prime example of a close relationship between doctor and emperor comes from the fourth century: that of Oribasius with the emperor Julian. He became a close and trusted friend of the emperor, who discussed with him matters of both private and public significance. Moreover, he played an important role in Julian's accession, according to Eunapius, whose language suggests that his power was all the greater in that it was indefinable.[9] Oribasius

[7] Crito: *F. Gr. Hist.* 200 (iiA. 931–2). Later, the son of a doctor, L. Gellius Maximus, was a legate and even a contender for the Principate (Dio 80. 7. 1). F. Millar, *The Emperor in the Roman World* (London, 1977) 491 ff., stresses the intellectual status of the doctors granted privileges by emperors. He links them with other intellectual groups granted privileges, such as orators and scholars. He sees similar values in the qualifications of those chosen as imperial secretaries and those who attended the imperial family as doctors and teachers.

[8] M. McCrum and A. G. Woodhead, *Select Documents of the Principates of the Flavian Emperors* (Cambridge, 1961), 458.

[9] Julian, *Ep. ad Ath.* 277 c; *Ep.* 4, Eunapius fr. 21. 2, = *Vit. Soph.* 498. R. C. Blockley comments (*The Fragmentary Classicising Historians of the Later Roman Empire*, ii (ARCA Classical and Medieval Texts, Papers and Monographs, 10; Liverpool, 1983), 133 n. 41) on the vague language used by Eunapius in this last passage to describe Oribasius' role in the accession plot. 'Oribasius . . . had many other virtues so outstanding that he actually made Julian emperor.' He suggests that Oribasius may have interpreted a dream of Julian concerning the overthrow of Constantius (see Julian, *Ep.* 4).

would thus represent a position of wide-ranging personal influence, arising not only out of his medical skills *per se*, but also from his learning in general. Of course, Seneca's pleasure that a doctor 'was worried on my account and not for his professional reputation' (*De Ben.* 6. 16. 4) would appeal to emperors also.

Personal services to the emperor in the medical and other spheres were often the spur for the granting of privileges to whole classes.[10] It may, however, be more than a coincidence that the major privileges granted to doctors up to the time of Vespasian's edict are connected with powerful figures who were all, in various ways, personally indebted to doctors. Dio represents the granting of immunities to doctors under Augustus[11] as resulting directly from Antonius Musa's saving of the emperor's life. The historian's language emphasizes the seriousness of the illness and Augustus' own part in the subsequent bestowing of rewards: besides the political advantages granted by emperor and senate, there was a personal financial reward from the emperor himself, as well as from the senate. Earlier, the citizenship had been granted to doctors at Rome by Julius Caesar,[12] apparently as part of a general move to encourage intellectual accomplishment and reward public service. We are not told, but it is interesting to speculate, whether Caesar's own history of ill health gave him a personal interest in this particular piece of legislation. Later, Vespasian's edict expatiated on the value of doctors to the good of the community. Yet, if he was not personally indebted to one particular doctor on medical grounds, doctors in Alexandria had given his power a boost in a different way, when they enhanced his authority in the East at a critical time by authenticating his 'miracles' (Tac. *Hist.* 4. 81). A notorious example of the lengths to which personal indebtedness of the powerful could go was Claudius' grant of immunity to Cos, which was, according to Tacitus, *Ann.* 12. 61, an undisguised personal favour to a single doctor, Q. Stertinius Xenophon.

[10] See Millar (n. 7), 491 ff.

[11] Augustus: Dio 53. 30. 3.

[12] Caesar and the citizenship: Suet. *Caes.* 42. 1. According to Plutarch (*Caes.* 17), Caesar was subject to headaches and fits, although he still lived an energetic life and fought off his weaknesses; Suetonius (*Caes.* 45) puts less emphasis on this illness. However, it may be significant that, of the three members of his household Caesar kept with him when captured by pirates on the way from Rhodes, one was a physician (ibid. 4).

The significance of this infiltration by doctors into the highest circles of power did not escape Pliny. With regard to Augustus and Antonius Musa, he simply acknowledges that Musa saved the emperor's life (29. 6–7). The rest of the passage, however, concentrates on the vast amounts of money doctors acquired from their imperial clients. It is clear that, in his view, their primary concern was mercenary,[13] not a disinterested concern for the patient's well-being as the Seneca passage (*De Ben.* 6. 16. 3) quoted above suggests. The problem seems to have been particularly acute under Claudius: the boast of Q. Stertinius Xenophon about his income, the fortune 'showered on' his brother by Claudius, the wealth amassed by one Arruntius, and Vettius Valens' abuse of his position by adultery with Messalina are all features of this reign, that is, of Pliny's own time (29. 7 ff.). The implication is clear: the emperor was unscrupulously exploited by those in whom he misguidedly placed his trust. Pliny could, however, have been more scathing in his criticism of Claudius himself here. Perhaps he made allowances for the problems of the strains of power (see above). Instead, his most outspoken attacks on the attitudes of doctor and patient are reserved for the upper classes of imperial Rome, who now have fewer responsibilities and whose disgraceful dependence on doctors is bred of idleness and luxurious living rather than the pressures of duty.

The question of dependence or self-sufficiency, integral to Pliny's thought in general, is of particular relevance to Roman attitudes to medicine. It has often been suggested that a strong do-it-yourself attitude existed since, before the introduction of Greek ideas, remedies were in the hands of the *paterfamilias* as an aspect of the upkeep of the farm and the family. That Roman native medicine was centred on the farm is of itself important to Pliny's ideology. Cato's notebook of remedies, by means of which he prolonged his own and his wife's lives and treated his son, slaves, and household, is held up by Pliny in opposition to the so-called art of the Greek doctors (29. 15–16). Moreover, he commences his main study of medicine in book 29 with two basic farm products, wool and eggs (29. 29 ff.). These, like the cabbage

[13] Other sources of evidence for this are given by V. Nutton, 'The Perils of Patriotism: Pliny and Roman Medicine', in *Science in the Early Roman Empire*, 30–58, who, however, also points out that wealth was not generally a feature of those members of the profession who worked in less exalted circles.

Cato swore by (*De Agr.* 156–8) were not merely 'natural' and uncomplicated: they were the fruits of a harmonious cultivation of nature by man, which made them the most familiar and easily obtainable of all Nature's cures. In this respect they are contrasted scathingly with cures from the phoenix's nest: 'remedies of life which appear at thousand-year intervals are a joke' (29. 29).

The general idea of self-sufficiency[14] is summed up well by another layman, the Roman encyclopaedist Celsus, in his treatise on medicine written in the reign of Tiberius. Ideally, the healthy man who is strong and his own master should not place himself under arbitrary rules or have the need for a doctor, but should give himself variety in life (*De Med.* 1. 1. 1). Some time after 81, Plutarch wrote his treatise on health-care. Significantly, it is not a technical work but commonsense advice, addressed by an informed layman to a professional physician. Plutarch draws the distinction between self-sufficiency and neglect (*De San. Tuenda* 26, 136 E). A man should know his own pulse and basic idiosyncrasies and not have to ask a doctor about such things.[15] Self-sufficiency is Pliny's main point in *HN* 22. 14 ff. Men are indifferent to health 'because we believe that it is the duty of others; that they make it their duty on instruction from us; and that doctors have made the necessary provision for us' (22. 15). He makes a rhetorical point out of man's zest for life in its pleasures and luxuries but neglect of its real interests. 'Never was there more covetousness and less care taken of life' (22. 14–15). 'We live in reliance on the good faith of others' (22. 15) encapsulates the opposite extreme of the self-sufficiency ideal. The contrast is heightened by the doctors' arrogant and irresponsible behaviour in their clients' households (29. 23), 'their myriad ways of changing their minds, their orders to the kitchens'.[16] In *HN*

[14] General health-care in Celsus and Plutarch: see also below on Greek dietetic medicine.

[15] The incident behind Plutarch's observation is recounted by Tac. *Ann.* 6. 46. Tacitus' point is ironic: Tiberius is sneering at doctors, when in fact he is defying his failing health with *libidines*. For the balance between neglect and hypochondria, cf. Marcus Aurelius, *Med.* 1. 16. 5.

[16] Seneca in *De Brev. Vit.* criticizes those who clasp the knees of physicians in supplication if they fall ill (8. 2), and those who have everything done for them (12. 6); cf. Pliny, *HN* 29. 1, above, Ch. VI, intro. Marcus Aurelius (*Med.* 9. 41) quotes with approval Epicurus: 'In my illness . . . I did not let physicians be

28. 53, on the other hand, remedies such as exercise and fasting, which demand positive action and will-power on the part of the patient, are approved. 'It would be all the less seemly to pass over the remedies which rely on the individual's will-power, such as abstaining from food or drink . . . which is considered one of the most outstanding cures.'

Finally, it is worth recalling Pliny's comments on those beliefs which encouraged a passive attitude (Ch. III. 1, p. 95). The image of the blind multitude hurling itself into dependence, now upon *Fortuna* and now upon astrology, bears a resemblance to the description of the arbitrary changes of theory made by the doctors and adopted perforce by their blindly dependent patients. Remedy succeeds remedy in fashionable circles: *mutatur ars cottidie* (29. 11); and 'no art has been more fickle nor is changed more frequently even today . . .' (29. 2). Medicine is as changeable as *Fortuna* herself, yet doctors are allowed to rule men's fates (*hi regebant fata*) as the stars do in astrology (29. 10). Perhaps actual associations of astrological lore with medicine were in Pliny's mind when he wrote this: in the same passage, he mentions Crinas of Massilia (29.9), who regulated diet according to movements of the stars.[17]

B. PRESERVATION OR PAMPERING? THE RICH HYPOCHONDRIAC

There was nothing in the receiving of medical treatment by the rich that was in itself contrary to Pliny's views. Philosophers do not consider health a bad thing. Even the Stoics said it was to be preferred (Sen. *Vit. Beat.* 22.2 ff.). Pliny himself, although he puts forward the view that death is a blessing, can castigate those who do not look after their own health: 'I am amazed to see neglected the means of preserving health, of banishing physical

insolent as if they were doing great things, but my life went on well and happily.' Whether or not the doctor was actually a slave, the feeling was that he should be a subordinate, leaving his patient in control of himself. On this attitude to slaves in general, see 28. 56: L. Lucullus would not exercise moderation in eating, so had a slave enforce control on him. He thus suffered the disgrace of having his hands kept away from food even at public banquets, but incurred additional shame by the fact that he obeyed his own servant more easily than himself.

[17] Later, astrology was of some importance in Galen's medicine (ix. 910–13 Kühn).

pain, of warding off death' (22. 14). The concepts of basic health and its attainment by medicine were, however, not so much ignored as misinterpreted. Too much attention to health was as dangerous as too little. Plutarch (*De San. Tuenda* 24, 135 D), after castigating Epicurean withdrawal from public life, since it did not produce the perfect health of which they boasted, suggested that such an ideal was in any case unnatural. 'We ought . . . to preserve the natural constitution of our bodies, recognizing that every life has room for both disease and health. Illness did not bar a man from an active and useful life (*De San. Tuenda* 8, 126 C, see below, this section). Moreover, such an idle, passive life encouraged indulgence. In the case of rich invalids, medicine might degenerate from its function as an art of necessity. A rich man would provide only such medical care as was necessary for his slaves, and this in the most economically efficient way. However, he was less likely to apply such objective and practical criteria to himself. He could even become a hypochondriac, pandered to by remedies that were not so much more efficacious as more luxurious and expensive.[18] In an anecdote about the emperor Commodus, Galen says that it is customary to treat emperors with the greatest care, and a remedy using expensive nard, rather than the more ususal spiced wine, is proposed (xiv. 659 Kühn). Elsewhere, Galen is at pains to distinguish between purely cosmetic preparations, which imperial ladies insist on having, and medical preparations which protect natural beauty throughout the body (xii. 434 Kühn).

The use of luxurious treatments is of immediate relevance to the theme of misplaced values discussed in the last chapter and elsewhere (Chs. I. V. 2. D). The rich come to believe that nothing is of any use unless it is expensive (29. 28, Ch. V. 2. D), a misapprehension which has affected other classes in other ages.[19] In

[18] For rich hypochondriacs, see Dio Chrysostom's description (*Or.* 52. 1) of the gentle routine of rest and light exercise for an invalid. But see also below, n. 26.

[19] Note the doctor-poet George Crabbe's experience in 18th c. Aldborough: '. . . his ignorant patients, seeing him return from his walks with handfuls of weeds, decided that, as Dr. Crabbe got his medicines in the ditches, he could have little claim for payment. On the other hand, he had many poor relations, and some of these, old women, were daily visitors, to request "something from comfortable cousin George"; that is to say, doses of the most expensive tonics in his possession' (*The Life of George Crabbe by his Son*, ed. E. Blunden (London, 1947), 30).

HN 24.4, the particular importance of this in the context of medicine is to be seen. Pliny talks of drug factories in which 'each man is promised a new lease of his own life—at a price'. Life itself, not just non-essential luxuries, has become the subject of monetary evaluation. It was a commonplace among moralists that the necessities of life were always cheap and simple.[20] Pliny hints that there is an unnatural, magical quality to ingenious compounds. 'Compound prescriptions and mysterious mixtures are chanted (*decantantur*) . . .' (24.4). Real necessities were also easy to come by, and there was no need to search for them beyond the sea. 'These things alone had Nature declared remedies, found everywhere, easy to discover and costing nothing—the things by which we live' (24.4). In *HN* 22.118, Pliny says that, in any case, plants from such far-flung places as India or Arabia (24.5) are unsatisfactory as drugs— why else should the locals sell them?—though they may suffice as perfumes and unguents. This does not tally with his praise of the *Pax Romana* for making the Empire the framework of a world-wide commerce in healing plants, however (27.2–3, Ch. V.2.c). It is the first hint encountered so far of the problems Pliny has in making too clean a contrast between Greek medicine and Roman herbals.

Another aspect of the danger of luxury in medicine was its contribution to that passive outlook in life discussed in the previous section. *Luxuria* in the form of high living, and medicine, could form a vicious circle. Plutarch remarks (*De San. Tuenda* 8, 126c) that contempt for health is least profitable for those who make pleasure their chief aim, though a man can be a philosopher, general, or king even if he is diseased. Pliny is not slow to point out the iniquity of those remedies which created a greater capacity for pleasure, such as hot baths to aid the digestion (29.26) and emetics which allow further heavy drinking. Indeed, 'these are the things which have destroyed the morals of our empire: the practices to which we submit when healthy . . .' (ibid.). The examples listed in this passage suggest a facet of Hellenistic medicine which Pliny might well see as corrupting, after observing its misinterpretation by his contemporaries. This is the idea that medicine is not simply the science of curing a specific disease when it occurs, but also involves keeping a

[20] Hor. *Sat.* 1.6, 2.2 are good examples.

constant check on the health. This dietetic medicine had been condemned by Plato for disrupting the workings of the state (*Rep.* 3. 407 C–D).[21] With the curtailing of independent political activity, the temptation to become addicted to these practices must have been even greater for the well-to-do Roman of the Imperial era. To one like Pliny, who had no trouble in employing himself usefully, even when not involved in state affairs, such behaviour was still unworthy of a Roman. That he is indeed referring to dietetic theories is suggested by *HN* 29. 26: the depilatory, digestive, and emetic practices of the healthy which he is describing are openly claimed as medical methods by leading doctors. Plutarch, although recommending a kind of regimen, makes it as natural as possible, a simple life-style with all-important flexibility (*De San. Tuenda.* 26, 137 B). He warns against becoming enslaved, even to a non-indulgent regime (23, 135 A). The illnesses he stresses arise from excessive pleasure: laziness is bad for health (24, 135 B ff.) and natural relaxation should not be self-indulgence but a change which is neither pleasurable nor painful (25. 136). If he can thus address 'scholars and men in public life', who may overwork as well as indulge unwisely (27, 137 C ff.), how much greater must the risks be for those who have time on their hands?

Sometimes the blind following of novelty and fashion led to habits whose appeal to luxury was not even very obvious. Pliny says that people had become so enslaved to ideas of health suggested by doctors that cold baths[22] enjoyed popularity because of their novelty: 'We have seen old men, of consular rank, actually going blue with cold in order to show off!' (29. 10–11). Plutarch is equally scornful of a practice which is 'more ostentatious and juvenile than healthy' (*De San. Tuenda* 17, 131 B).

The rich,[23] then, were especially vulnerable to unscrupulous treatment by the medical profession. They alone had the time and money to take complete care of their health, and were the

[21] See E. J. and L. Edelstein, *Asclepius* (Baltimore, 1945), ii. 178–9.

[22] Sen. *Ep.* 53. 3, 83. 5.

[23] Plutarch does mention the rustics, whose continual bustle of work exacerbates their bodily ailments (27, 137 c). For them, necessity rather than leisure would produce a certain preoccupation with medicine. Those doctors who did public medical 'displays' must also have relied on the interest of the *vulgus* in any kind of *spectaculum*, an entertaining discourse of skills and marvels and practical demonstration. Cf. Dio Chrys. 33. 6; Plut. *Adulator* 32, 71a.

natural subjects for dietetic medicine. In one sense, therefore, the physician could do more for the patient the richer he was. Traces of this attitude may be reflected in Galen's remark about the utmost care being given to emperors, quoted earlier. More often, however, the rich must have been natural patients for the less creditable practitioners, as Galen himself suggests elsewhere: those doctors who seem, rather than are, learned are the type who flatter and visit the rich and powerful (xiv. 600 Kühn). Yet, despite his prejudice, it is clearly the patients, as well as the doctors, who are to blame in Pliny's view, because of the incorrect attitudes which they hold with regard to life and its values.

In this respect, it is, perhaps, ironic that the educated interest in medicine which inspired writings such as those of Pliny and then Plutarch may itself have encouraged a tendency to hypochondria. In later times, Marcus Aurelius' tutor, Fronto, who frequently consulted doctors, was not personally ignorant of medicine, but excessively interested in his own ailments.[24] The highly educated Aelius Aristides enjoyed perplexing the doctors with the complexities of his illnesses.[25] This preoccupation used to be assigned to a later age than Pliny's. In fact, Pliny's own testimony suggests that there was a gradual development and hypochondria was not attributable solely to the circumstances of the later second century and after.[26] The morbid preoccupation

[24] e.g. *Ad M. Caes.* 3. 8. 3; 5. 21, 27, 29, 33. Marcus commiserates (e.g. 5. 28, 30, 31, 34). Gout was a fashionable ailment (D. Gourevitch, *Le Triangle hippocratique dans le monde gréco-romain: Le malade, sa maladie et le médicin* (Rome, 1984), 217–47), though Fronto seems to have had genuine reason for complaint (Gellius 2. 26. 1, 19. 10. 1; Artemidorus 4. 22).

[25] See Aelius Aristides, *Sacred Tales* (ed. Keil), 40. 3. For references, J. Scarborough, *Roman Medicine* (London, 1969), 103 ff.

[26] See E. R. Dodds, *Pagan and Christian in an Age of Anxiety* (Cambridge, 1965), who saw valetudinarianism as part of a search for security in a less rational era. Cf. A.-J. Festugière, *Personal Religion among the Greeks* (Berkeley, Calif., 1954), 86 ff., and G. W. Bowersock, *Greek Sophists in the Roman Empire* (Oxford, 1969), 59–75. However, it is now generally agreed that the tendency to morbid interest in illness was common to antiquity generally and not confined to a particular age. W. H. S. Jones's theory ('Ancient Documents and Contemporary Life', in E. A. Underwood (ed.), *Science, Medicine and History: Essays in Honour of Charles Singer* (London, 1953), i. 100–10) attributing ancient hypochondria entirely to fear of malaria has rightly been regarded as an overstatement, but does at least point to a secondary factor—fear of certain diseases—in the conditioning of ancient attitudes to sickness, which is not confined to a particular era (see P. A. Brunt, *Italian Manpower* (Oxford, repr. 1987) App.

of Fronto or Aristides might seem far from the sturdy pragmatic tones of Pliny's exhortations to his fellow countrymen to be self-sufficient and learn the basic medicines supplied by *Natura*. Yet it might be argued that his enthusiasm for such remedies also betrays his own form of hypochondria. The part of the *HN* (bks. 20–32) dedicated primarily to the description of medicines is disproportionately large, no fewer than twelve books out of thirty-seven. Remedies feature to a lesser extent in many of the other books. Rather than contradicting the idea of Pliny as a no-nonsense advocate of an active and independent life, this interest in the things 'which preserve life' may rather be its natural corollary. It is an idea to which we shall return in the final section.

C. PROFIT VERSUS HUMANITARIANISM

We have seen that medicine could, in certain respects, lead to a corrupt way of life. Even with regard to the principles of the primacy and dignity of life itself, the ethics of scientific medicine were not always in accord with Pliny's humanitarian principles. Although a feeling that medicine should be subservient to the needs of the patient was increasing, even Galen, at a later date and with a command of philosophical ideas including later Stoicism, could imply that a doctor might not necessarily be motivated by humanitarian ideals. Although he says, on the authority of Plato, *Rep*. 340 E–347 A, that the doctor's job is care for the body's health, he adds that it may be for a variety of reasons that he practises medicine, such as honour, glory, or tax-exemption (*De Placitis* p. 763 Müller).[27] It has been suggested that the idea of the necessity of a humanitarian motive is rare and may be attributed to later Stoic ideas,[28] such as those of Cicero,

18, 611–24 and 726 for malaria as a perennial, if variable, presence in Italy). The 'preoccupation' which Jones found most pronounced in Plutarch, *De San. Tuenda*, is already clear in Pliny, to judge from his figures of 97 references to malarial fevers and 80 to related diseases in the *HN*. Conversely, if we wish to follow 'hypochondria' beyond the 2nd c., there is no shortage of evidence in the Christian writers of the 4th c., e.g. Basil, *Ep*. 162.

[27] Edelstein, 'The Ethics of the Greek Physician', *Papers* (n. 5), 336.

[28] Ibid. 344. Cicero is drawing on Panaetius. A poem of a 2nd-c. Stoic philosopher and doctor, Serapion, is inscribed on the wall of the temple of Asclepius at Athens, in which he says that it is the doctor's first duty to heal all men and be a brother to all.

De Off. 3.10.43, where, using the example of the judge, he says that all professions have their own ethos. It appears in the Roman physician Scribonius Largus, a younger contemporary of Celsus. Writing a little before Pliny under the patronage of the emperor Claudius, he talks of the necessity of having *humanitas* and *misericordia* in accordance with the *voluntas* of medicine itself (*Compositiones* pref. 3 Sconocchia). Largus' committal to all men indiscriminately is also interesting. 'Medicine does not grade men according to wealth or rank, but promises to give her aid to all who ask without making distinctions . . .' (ibid. 4). Curiously, Pliny appears not to have heard of Largus. Perhaps he is deliberately passed over for the sake of the contrast between Greek medicine and Roman herbals; Pliny will not admit that the medical profession includes Romans.

Pliny's own humanitarian stance shows an advance on Largus in some respects. Like other writers on medicine, he does fall into the trap of recording remedies he does not approve, for the sake of possible success, or where there is no alternative. He is more unusual in expressing disapproval of some remedies because they are crimes against humanity and not just because they are magical (see Ch. III.2.A.iii). When remedies are 'crimes', Pliny refuses to discuss them, even when there is no alternative and in spite of the fact that he does appear to believe in their efficacy. Thus, while he will, occasionally, record without criticism magical remedies where everything else is useless (29.98), the drinking of gladiators' blood by epileptics is condemned, 'as if it could possibly be considered healthy for a man to turn into a beast' (28.5), although his language implies that it might work. 'Those who use such remedies, will none the less die, even though [they] may have lived (longer) through disgusting and wicked practices' (*etiam cum obscaenus vixeris aut nefandus*, 28.9; tr. Jones). In a similar passage of Largus, there is not the same degree of condemnation. Although the latter admits that a remedy from gladiators' liver 'falls outside the medical profession, though some have found it helpful' (*Comp*. 17), he includes it on grounds of empirical evidence.

Another divergence between Pliny's general ideals and those found in Hellenistic medicine occurs over the question of abortion. We saw, in Ch. III.2.A.ii, that this could be portrayed by Pliny as being magical in that it disrupted the normal processes of nature. *HN* 25. 24, with its reference to poisons as well as abortives, may also be a reference to the ignoring by Greek doctors of the ideals which in earlier days had been enshrined in the Hippocratic Oath:[29] 'I myself am amazed that even harmful plants have been described by the Greeks and not only poisons (since the condition of life is such that death has always been a refuge even for the best of men). But what excuse was there to indicate the means of causing madness, abortion, and other similar effects? I myself am mentioning neither abortives nor any other magical monstrosities.' The doctors are to be blamed, not for pointing out suicide poisons, but for drugs which disturb man's natural state of living. Pliny goes on to say that he will not even mention love-potions, since these can cause unintentional death. It is enough to point out plants healthful to life, and discovered in order to preserve it (25. 25).

Abortives and contraceptives were among the remedies readily accepted from Hellenistic medicine by Rome.[30] Thus, there is some excuse for Pliny's anti-Greek slant in the comments in *HN* 25.24. However, it is clear that 'native' methods of a greater or lesser efficacy existed. Pliny tells us that it was believed that lamp-smoke could cause *abortus*, as could sneezing (7. 42–3) and eating wild cucumber (20. 9).[31] In each case, Pliny is describing a fortuitous result, but such knowledge could be put to deliberate use. Thus, even if the most reliable methods 'were limited in the empire to those who commanded the services of a Greek physician',[32] the desire, at least, was present in folk medicine.

To assess Pliny's objections to abortion in the wider context of ancient views on the subject is not easy. It has been suggested that the Hippocratic Oath itself arose from Pythagorean doctrine, which was exceptional in being the only philosophic school to hold that life was present in the foetus from conception.[33] Yet it

[29] It is very unclear who took the oath and how representative its ideals were in antiquity. See L. Edelstein, *The Hippocratic Oath* (Baltimore, 1943) and G. E. R. Lloyd, *Hippocratic Writings* (London, 1978), 20.

[30] K. Hopkins, 'Contraception in the Roman Empire', *Comparative Studies in Society and History* (1965–6), 124–51; Scarborough (n. 25), 101–2 and nn.

[31] *HN* 30. 130 records a 'Magian' belief of the same type: stepping over a raven's egg will cause abortion.

[32] Scarborough (n. 25), 101.

[33] Edelstein, 'The Hippocratic Oath', *Papers* (n. 5), 3–63, esp. 9–20 on abortion.

is not correct to see the ancient attitudes to abortion as being necessarily based on the question of foetal life.[34] The Stoics believed that life was present in the foetus only from birth.[35] However, the later humanitarian trend of Stoicism, combined with the implication in Aristotle (*Pol.* 7. 1335^{b} 22–4) that animal life was attained some time before birth, may have led Largus to voice doubts on this. His phrase is vague—'the uncertain promise of a human being' (*Comp.*, pref. 5)—but he uses it to enlarge the horror of killing a grown man who is indubitably a live human being. While Pliny's condemnation seems emphatic in comparison, however, it does not follow that this was due to a belief in the foetus' claim to human life. There was a considerable undercurrent of much vaguer moral disapprobation in Roman thought, which classed abortion with breaches of decorum in the case of the upper classes. Such feelings would have been intensified by Augustus' attempts to increase the birth-rate. Later, Seneca writes in *Ad Helv.* 16. 3–4 in tones of approval that the addressee 'did not abort the promise of children conceived within her womb', to which he adds such social virtues as lack of make-up, modest clothing, etc. A similar way of thinking is to be seen in Plutarch. 'Purges are used . . . after the manner of licentious women who employ drugs and instruments to produce abortion for the sake of the enjoyment of conceiving again' (*De San. Tuenda* 22, 134 F).

If the statement in *HN* 25. 24 does not indicate a belief in pre-natal life, however, there is still every reason to think that Pliny was motivated at least by a belief in not interfering with the processes of nature. The context, of maddening poisons and of *magica portenta* (see Ch. III. 2. A. ii), supports this idea of unnaturalness. The distinguished Methodist physician Soranus, who practised in the principates of Trajan and Hadrian, reported a medical controversy in his time over the use of abortives. The doctors who opposed their use cited the Hippocratic Oath and said it was the duty of a doctor 'to safeguard and preserve

[34] S. Dickison, 'Abortion in Antiquity', *Arethusa*, 6 (1973), 159–66, suggests that this only assumed its prime position as a criterion in Christian times. Other reasons are also present earlier: the first rescript against abortion under Septimius Severus (*Digest* 48. 19. 39; 48. 8. 8; 47. 11. 4) was concerned with infringement of the husband's rights if a woman, after divorce, had an abortion against his will.

[35] *SVF* ii. 756.

what Nature has created' (*Gyn.* 1.60). Such an idea is perhaps also reflected in Pliny's figurative use of *abortus* in the sense of 'arrested growth' generally. In fact, the majority of the attested uses of this meaning of *abortus* appear in the *HN*.[36] Pliny applies it to both plants and animals. In all cases, it denotes unnatural curtailment of the natural development of fruit or offspring. Sometimes, this is just a freak of nature, 'miscarriage': oysters, when frightened, produce *physemata*, wind-pearls (9.108). Fig-trees occasionally produce fruit which never ripen (16.95) and bees make a bitter kind of wax if they are infertile, diseased, or lazy (11.50). Diseased oats, too, are said to experience a kind of *abortus* (18.150). In the case of dwarf trees, however, the *abortus* is caused by man's interference. This deliberately unnatural 'abortion' is spoken of in terms of disapproval by Pliny (see Ch. II.5).

In *HN* 29.85, Pliny describes an amulet which may be hung around women's necks to stop them conceiving. He says it is right for him to mention this one alone out of all such preventatives. It may be significant that his exception to his general rule in *HN* 25.24 is apparently contraceptive rather than abortifacient. Even if a concern for foetal life as motivation for the choice is dismissed as before, a preference for the prevention rather than termination of pregnancy might be postulated on the basis of the 'natural cycle' idea described in the last paragraph. That is, man might do less violence to nature by preventing a natural process altogether rather than cutting it off once it had started. However, the difference between abortives and contraceptives was often unclear in Greek and Latin terminology.[37] Elsewhere, Pliny uses *abortivum* of what is almost certainly another contraceptive (24.18).[38] In fact, the choice may be primarily on the grounds of safety. Pliny says (29.85) that his concern is for women ex-

[36] See *TLL* i, sv. *abortus*. Pliny even uses it in a light-hearted reference to the frustrated work of his literary critics (pref. 28).

[37] See e.g. K. Hopkins, 'A Textual Emendation in a Fragment of Musonius Rufus: A Note on Contraception', *CQ*, NS 15 (1965), 72–4. Besides Pliny's use of *abortivum* (n.38), the Greek term which he transcribes in 29.85, *atokion*, could bear either meaning: thus in Hopkins's emendation of Musonius fr. 15*a* it means 'contraceptive', but Scarborough (n. 25) points to Dioscurides 11.59.3, where *atokion* is applied after coitus.

[38] *HN* 24.18. 'There is a bizarre rumour (*portentum*) abroad that to rub cedar pitch on the male parts before intercourse prevents conception'. Cedar-pitch is also referred to in other classical references to contraceptives: Aristotle, *HA* 7.3,

hausted by child-bearing. Safety is a main criterion in Pliny's criticism of drugs and remedies generally, as seen, for example, in his concern about remedies that may damage the stomach (25. 41, 26. 58, 25. 154). It is also important in *HN* 25. 24, where emphasis is put on the danger of poisons and the accidental death caused by a love-philtre. The inclusion of abortives here suggests that they too were often dangerous to the recipient. Significantly, Caracalla's later rescript (*Digest* 48. 19. 38. 5) contained the same coupling of abortive and love-potion. Its concern, too, was the possibility of injury or death to the recipient, and, in cases of fatality, the death penalty might be imposed on the administrator of the drug.

It has been suggested in this section that Pliny put more emphasis on humanitarian ideals than was to be found in Greek scientific medicine of his time or earlier. His attack on the Greek medical profession may have some justification in this respect, since doctors had to make a livelihood out of medicine. Plato had implied that, unless there existed a spur, such as money-making (or as he prefers a penalty, as good men would not wish to be mercenary), the craftsman will have no incentive to help others (*Rep.* 1. 34 A). Certainly, if Pliny were looking back at the older Roman *medicina*, he might well consider that there had been a shift from the motive of personal necessity in the treatment of family and household to a more impersonal and mercenary outlook of the professionals. Whether the treatment of one's slaves was not also with mercenary motive was, of course, arguable. However, Pliny's outlook is to an extent justified by later evidence. Even in the fourth century, when doctors were being paid a subsistence allowance, they still had to be urged to 'minister to the poor honourably rather than to the rich shamefully'. In addition, there were blatantly commercial consequences of the professionalization of medicine at Rome. Epictetus comments '. . . as if a doctor were to advertise for patients, rather

583^{a} 22–4, where this, white lead, or oil can be applied to the womb and Soranus, *Gyn.* 1. 61. Pliny's account of this use of *cedria* follows that of Dioscurides 1. 77 (Wellmann) closely. The significance of Pliny's reaction, *portentum*, is perhaps alarm rather than disbelief *per se*. He has just listed the wide-ranging uses and the drastic results that this substance can have; preserving dead bodies but making living ones decay, rotting clothes, and killing animal life. (His verdict, that for these reasons he considers it inadvisable to use it as an indigestion remedy, sounds an understatement.)

than they come to him . . . though at Rome, doctors do advertise . . .' (*Discourses* 3.23.27).[39]

Yet, even if praiseworthy ethics did exist in the medical profession, Pliny is in no mood to look for them. The hero-god Asclepius could be connected with higher spiritual values than simple bodily health. Galen mentions Asclepius' correcting of the soul by the reproportioning of the emotions through the composition of songs and odes, or the practice of hunting (*De San. Tuenda* 1.8.19–21). According to Julian, the Muses 'train our souls with the help of Asclepius, Apollo, and Hermes' (*Contra Galilaeos* 235 B).[40] Pliny mentions the god only twice in any detail, both times maliciously, as the figurehead of a profession whose ideals are anything but spiritual. The early Romans refused to pay fees to medical profiteers, and for that reason they also kept Asclepius out of the city, the temple being on an island in the Tiber (29.16). In the unfavourable account of 'non-herbal' medicine (29.1 ff.), an old legend of the hero Asclepius is revived: he was struck by lightning for raising Tyndareus from the dead. Thus Pliny can say that the fame of the medical *ars* was increased by *crimen*![41]

2. Auctoritas *and* Ratio: *Doctors' Abuse of Power*

The official recognition of doctors as members of a profession at Rome[42] introduced an intermediary between the patient and the remedy. Perhaps this was, in itself, unnecessary in Pliny's view.

[39] Money: *Cod. Theod.* 13.3.8, *trans.* C. Pharr (New York, 1969). The date is 368. Such temptation must have been even more likely for the doctor who was not state-aided. Publicity: precedents existed, e.g. Menecrates of Syracuse, whose tactics are memorably described in Athenaeus 7.289 A–F.

[40] Edelstein, *Asclepius* (n. 217) i. 208–9, ii. 126–7.

[41] Ibid. ii. 46–50. The earliest version of the legend may have laid stress on a personal offence against the gods for the raising to life of specific heroes. The idea that, by raising the dead, he violated the laws of nature is the later explanation, though perhaps already implied by Pindar, *Pyth.* 3. This fits Pliny's general outlook, although he names a specific individual, Tyndareus. Cf. the variations in Ovid: *Fasti* 5.743–62 and *Met.* 15.531–6 (see *Asclepius* i, texts 75–6) suggest the breaking of natural law, while *Met.* 2.644–8 describes the raising of one individual against the gods' will.

[42] Pliny dates the establishment of Greek medicine at Rome to the arrival of Archagathus in 219 BC, on the authority of the 2nd-c.-BC annalist Cassius Hemina. What is important is the officializing of his position. There may have

His ideal appears to be that of a direct relationship between man and nature, in which nature gives cheap and simple herbal remedies to those willing to make the effort to find and identify them. It is certainly clear that he felt that the Greek professionals were entirely unscrupulous in their use of their positions of authority. The power and *benignitas* of *Natura* were being usurped by mere men with inflated ideas of their personal power. Their often questionable motives were encouraged by the desperation of the truly sick and the whims and hypochondria of some of the wealthy and leisured. We have seen that these suspect foreigners also attained a mobility of social position at Rome. Perhaps this ambiguity also enhanced Pliny's suspicions: were they respectable persons (*honesti*) or not? *Honesti* were worthier and more reliable than slaves with regard to the practice of agriculture (18. 21, see Ch. V. 1. A. i) He might equally well ask himself whether Greek doctors, with their foreign origin and ambiguous social status, were worthy of their position of power in the dispensing of remedies to their Roman conquerors. 'We are subject to foreigners and in this one art they actually rule their rulers' (24. 5). In the following sections, the abuse of their authority by doctors will be examined with particular reference to the power exercised by the possession of an *ars*. However, the idea of an antithesis between theory and practice will be brought into doubt once again as the true nature of Pliny's herbal 'alternative' is revealed. But first we return to the relation of the doctor's power to that of *Natura* herself.

A. THE ROLES OF MAN AND NATURE IN THE HEALING PROCESS

There had been tension between the roles played by man and nature in the healing process even in the early Greek medical writings. In a consideration of some works in the Hippocratic Corpus, Edelstein[43] concluded that, in the fight against illness, the knowledge of the physician was one factor, nature the other.

been Greek doctors in Rome before him, but he was given the citizenship and employed at public expense, which was normal in a Greek city but a unique occurrence at Rome so far as we know. See V. Nutton (n. 13), 38.

[43] Edelstein, 'Greek Medicine in its Relation to Religion and Magic', *Papers* (n. 5), 228–9 and n. 76.

The writers in question do not exaggerate the healing power of nature, in order to avoid the negativism which would result from strong belief in this. Later, Celsus does not claim too much power for the physician. It is not certain whether recovery is due to medicine, or to a sound body, or to good luck (*De Med.* 7, proem 1–2). However, when it was so uncertain why remedies worked, a cure would be claimed for the drug and thence for the physician who administered it, but failure could be blamed on the patient. As Pliny noted, with the prevalence of such theories as dietetics, the latter might well have made an error or omitted some element in his routine (29. 18, 'his lack of self-control is blamed'). Pliny's criticism of compound drugs implies that he recognized that doctors could hide behind the compounds they prescribed: one element might in fact work, but he would not know which one. Such mixtures would, moreover, impress the client and give an enhanced impression of the doctor's own authority. The ideal of the simplicity of Nature's remedies is lost. Indeed, Nature's power is wholly obscured by the doctors' clever deceits (24. 4).

It is possible that the remedies called 'sacred' may have originally been inspired by the divinity of plants arising from the divinity of nature.[44] It would be easy to imagine how these came to increase the prestige of those who administered them, rather than that of the plants themselves, or their true creator, *Natura*. Scribonius Largus repeats with approval the third-century-BC Alexandrian physician Herophilus' saying that plants are the hands of the gods (*Comp.*, pref. 1), but Galen rationalizes this to the role of the doctor: plants act like the hands of the gods, since it is efficacious for the man who uses them to be trained in logical methods and also to have a natural understanding (xii. 966 Kühn). Pliny would not have quarrelled with the rational use of the human intellect in this respect. It seems, however, that he suspects misuse of intellectual skills in the case of some contemporary Greek practitioners. Their motives are those of self-interest rather than logical reasoning: deceitful rhetoric rather than honest use of *ingenium* is behind the arbitrary and short-lived medical fashions. 'We are blown along by the hot air generated by Greek intellects and it is obvious that whichever of

[44] Ibid. 230 ff.

them is the most powerful speaker at once obtains supreme command over our life and death' (29. 11). This is the core of his distrust of Asclepiades of Bithynia, who practised medicine at Rome in the first century BC. Pliny has to admit that his enlightened treatments were often more effective as well as more pleasant than older ideas. However, he was, originally, a teacher of rhetoric who was dissatisfied with his earnings and 'suddenly transformed himself into a doctor' (26. 12). 'One thing alone moves me to anger, that one man of a very superficial race . . . in order to increase his income, suddenly gave to the human race rules for health which, however, have been generally discarded' (*HN* 26. 15).[45]

B. THE POWER OF PROFESSIONAL LEARNING: MISUSE OF *RATIO*?

This suggestion of a deceitful, false use of *ratio* in medicine is enhanced by another factor. It was precisely to give magic a greater intellectual standing as an *ars* that Ptolemy was to compare it to medicine (Ch. III. 2. A. i). Medicine's position as a 'parent' of magic is never far from Pliny's mind and increases his bias against theoretical medicine. It is worth reiterating briefly some of the conclusions on this relationship reached in Ch. III. Learned Greeks are castigated (30. 8) for seizing upon Osthanes' magical learning with such enthusiasm. 'It was this Osthanes more than anyone who raised the Greeks not merely to eager but to insane desire for his art.' It is true that, in *HN* 30. 10, Pliny remarks that Greek medicine and magic flourished separately at the same time: Hipparchus was expounding medicine when Democritus was expounding magic. Yet there remains a strong link in the *HN* between magic and medicine as practised by the Greeks. Greeks are specifically associated with the inhuman remedies from human bodies, adopting the barbaric ideas pro-

[45] E. Rawson, 'The Life and Death of Asclepiades of Bithynia', *CQ* 2 32 (1982), 358–70 (*Roman Culture and Society* (Oxford, 1991), 427–43) comments (pp. 365–436) on doctors' use of rhetoric and evidence of prejudice against this. For Asclepiades' dates and milieu, *passim*, reinforced by D. Gourevitch, 'Asclépiade de Bithynie dans Pline: Problèmes de chronologie', *Helm.* 37 (1986), 67–81. For a more detailed reconstruction of his theory from the ancient sources including Pliny, see J. T. Vallance, *The Lost Theory of Asclepiades of Bithynia* (Oxford, 1990).

mulgated by the Magi (28.6–7, see Ch. III.2.A.iii, cf. above, sect. 1.c). Again with regard to medical compounds, Pliny seems to regard their invention as a false use of ingenuity analogous to the magician's impudent attempts to 'improve' on the natural power already inherent in the natural ingredients (Ch. III.2.A.ii). Describing the Mithridatic antidote, he insists that human ingenuity could not have evolved such complex proportions. They are clearly 'an ostentatious and prodigiously boastful medical display' (29.25). By false ingenuity, doctors deceive their patients with regard to the true efficacy of their remedies, at the same time enhancing their power in the eyes of the credulous by an aura of mystery, another trick of the magician's (Ch. III.2.A.i, ii). Even perfectly rational medical deductions might cause astonishment and wonder among uninitiated laymen, a point not missed by Galen (xviii. 241 Kühn).

The language used to describe the quick changes in medical theory and the upstaging of one doctor by another is that of political intrigue—*factiones*, *adsectatores*—and even of an armed struggle for victory. In Pliny's account, the history of medicine is that of a series of rifts between *factiones*, as doctors revolt against the tenets of their predecessors. 'Another faction, calling themselves Empirics because they relied on experience, started in Sicily . . . These schools disagreed and Herophilus condemned them all . . . Then his sect also was abandoned. . . . Valens, gaining followers (*adsectatores*) and power, started a new faction' (29.6–8). Thessalus of Tralles, founder of the Methodist school, actually recorded his 'victory' in the contest by having the epithet *iatronices*, 'the conqueror of doctors', inscribed on his tombstone (29.9). Whichever doctor won the war of words became *imperator* of life and death (29.11, see end of last section). After recording Thessalus' victory inscription, Pliny describes him as outstripping in popularity any charioteer or actor. This suggests that he is thinking not only of the *factiones* of political struggles but also of the *factiones* of the Circus or stage. There too, the monetary rewards were great for the celebrities. For Pliny, the art of medicine is as disreputable and as unedifying as such popular entertainment; it is nothing but showy treatments and ready words, mere theatricality.

By means of such language, Pliny ingeniously suggests the falseness of the *ars medicinae*. The changes are irrational in their suddenness and violence. This is particularly so in the case of Thessalus. 'The same generation in the principate of Nero *rushed over* to Thessalus, who *swept away* all received doctrines and preached against doctors of every age with a sort of *mad frenzy*.' Pliny ironically suggests that his victory inscription is a fair estimate of his *prudentia* and *ingenium* (29. 9). The radical and frequent nature of the changes in medical theory also suggests that this was no true body of learning, capable of being passed on and added to by posterity in accordance with Pliny's ideal of human progress (see Ch. II. 1 and below, sect. 2. c).

Is Pliny being fair? We have already seen that he is biased by the Greek domination of the medical profession. This in itself produces a distorted picture of the profession. Those at the humbler end of the scale, working among the masses, are ignored. Yet it is unlikely that they fulfilled Pliny's idealistic conception of herbal medicine even if they were Roman. Passages in the *HN* itself describe how native herbalists (*nostri herbarii*, 25. 174) can sometimes adopt the criminal habit of replanting a herb which has cured a patient in order to make him ill again. Upon such revitalization, the healing powers extracted from the plucked plant were supposedly drawn back into it and out of the recipient. This seems indistinguishable from the noxious and magical practices which Pliny ascribes to the Magi.[46]

The manner of Pliny's arguments against the Greeks is also disingenuous. Practical treatments are given with no indication of the theory behind them. This makes it all the easier for Pliny to insinuate that they are immoral, degenerate, effeminate, etc. The difference between the medical sects interests him only in so far

[46] See also 21. 144; 26. 24. Even allowing for the disreputable image of doctors generally found in rhetoric, Cicero's discussion of the country-town scandal in *Pro Cluentio* suggests the native doctor was not an idealistic herbal healer. The presumably Roman L. Clodius, *Anconitanus quidam*, is called *pharmacopola circumforaneus*, with its overtones of quackery and unscrupulous sales tactics. While his surroundings might suggest that herbal remedies were the medicines most readily available to him, he probably got hold of anything he could (*Pro Clu*. 40). In *Pro Clu*. 47, a Greek doctor is mentioned, but with a slave who is *integer*. This contrasts with an apparently Roman doctor, A. Rupilius. He has a Greek slave, Strato, who (§ 179) murdered and stole and was eventually given a well-stocked medical store of his own. No doubt, he, too, was not unduly worried whether his drugs were local or imported. While it is possible that Clodius and Rupilius were of Greek origin, the tone of the passage does not suggest that their unscrupulous dealings were attributable to their race.

as it reinforces his arguments against the coherence and credibility of medical theory.[47] Unlike Pliny, Celsus had considered the relation of theory and practice and reached a suitably Roman compromise. Cause, he concluded, varies, and is uncertain, as opposed to the certainties of experience, but deeper research made Erasistratus and Herophilus not *medici* merely, but *maiores medici* (De Med. i, proem, 47–8). The art of the physician ought to be rational, but should also draw instruction from evident causes, rejecting obscure ones from medical practice but not from medical research (ibid. 74).

Pliny approved the combination of *ratio* with *usus* in such practical arts as farming (Ch. V. 1. B. i). Why does he seem loath to admit the same compromise in medicine? After Herophilus, 'experience (*usus*), the most efficient teacher of all things, and especially of medicine, degenerated into words' (26. 11). Why is *usus* 'especially' important in medicine? Elsewhere in the *HN*, Pliny suggested that the practical ideal was the combination with *ratio*, but here *usus* alone is assigned to the favoured herbal medicine and *ratio* alone to the Greek medicine, the two being mutually exclusive. The *scholae* of the sects are contrasted with *solitudines*, the natural habitat of herbal remedies. It is time to examine this ideal of herbal medicine, to see how far it encapsulates Pliny's ideology and stands in opposition to the principles of Greek medicine. We need to determine whether Pliny's ideals, combined with his Roman patriotism, account adequately for his stance, or whether there may not be other factors, not considered so far, to be taken into account.

C. HERBAL MEDICINE AND GREEK 'THEORY': PROBLEMS OF CLASSIFICATION

By being too dismissive of Greek medicine's theory, whether pure or applied, and too insistent on herbal medicine's tradition and practicality, Pliny creates problems for himself. The opening paragraphs of *HN* 25 imply that in fact herbal medicine requires

[47] It should, however, be remembered that Galen was later to talk in terms of competition with other doctors, and to complain that some chose the sects they belonged to haphazardly, merely because a relative or friend belonged (see O. Temkin, *Galenism: Rise and Fall of a Medical Philosophy* (Ithaca, NY, and London, 1973), Ch. I for references).

intelligence and rational procedure and even some outstanding human minds to give it shape as a viable and transmittable body of knowledge. We saw, in Ch. III. 2. A. iv, that lack of systematic means of transmission put the propagation of herbal knowledge at a considerable disadvantage in comparison to the learned writings of the Magi. Of particular relevance to the present study of Plinian attitudes to medicine is a section of the passage, not so far considered in this context, in which Pliny praises Mithridates (25. 5–7, cf. III. 2. A. iv). Doubts about the validity of Pliny's division between Greek and herbal medicine appear at once: Mithridates was the creator of an antidote which surpassed all others in complexity and number of ingredients, earning it particular scorn and condemnation from Pliny elsewhere (29. 24–5, see Ch. II. 2). In the present passage, however, the king is praised for his use of *ingenium*. In fact, he epitomizes the Plinian ideal. He blended practical investigation with the ability to communicate his discoveries in writing. There was nothing random or mysterious about his experimentation; it followed a carefully devised theory. 'He alone worked out the method of drinking poison daily after first taking antidotes in order to render it harmless by sheer custom. He had a wide-ranging and brilliant mind and was particularly interested in medicine' (25. 6). His was no mindless amassing of traditional ideas; his discoveries were systematically recorded and preserved. 'He collected detailed knowledge from all his subjects . . . and left among his possessions a bookcase of these treatises with specimens and their properties' (25. 7).

There is implicit recognition here of the fact that nature provides the basic ingredients of even the most complex compound medicines.[48] In this passage, at least, Pliny does not

[48] G. E. R. Lloyd, *Science, Folklore and Ideology* (Cambridge, 1983), as part of his thesis that 'much of Greek science consists in the rationalisation of popular belief' (207), discusses the overlap between the Hippocratic pharmacologists and the 'root-cutters', a relationship which bears similarities to that of 'Greek medicine' and 'native cures' in Pliny. Lloyd stresses the similarities, noting that the Hippocratic doctors often use the same herbs as the root-diggers. Although they try to contrast themselves with the latter, paying less attention to popular beliefs and practices, the difference was 'more a matter of presentation than of substance' (134). Cf. Edelstein, 'Greek Medicine as Science and Craft', *Papers* (n. 5), 140, who cautions against thinking of the physician and the traditional leech as distinct classes: '. . . we may imagine a broad stratum of practical men between the leaders whose ideas the [Greek medical] literature reflects, and the practitioner bound by mere tradition' (152).

consider the Mithridatic antidote overingenious (see 29. 24–5) to the point of irrational absurdity. Instead, its inventor was, 'as we know by evidence besides report, a more industrious investigator of life than any of those who came before him'. An antidote to poison is, of course, an unambiguous aid to the promotion of 'life', which may be the consideration uppermost in Pliny's mind at this point (cf. 25. 7, Ch. III. 2. A. iv). However, it is impossible to ignore the tenor of the passage as a whole, stressing as it does the correctness of Mithridates' principles in the acquisition and transmission of knowledge and the importance of his intellectual capacity in this respect. It should be noted that Pliny also praises another aspect of his intelligence here: the ability to speak all the languages of his vast empire, enabling him to address all his subjects personally. It is another example of the practical use of *ingenium*, which is also in accordance with another Plinian ideal, universal communication. We are reminded once again of the world-wide commerce in healing herbs masterminded by Rome (27. 2–3).

The Mithridates passage thus belies the strict Greek–Roman, theory–practice division of medicine which Pliny advocates in his more polemical passages. Here it is suggested that theory and practice can be blended in the exploitation by man of *remedia* provided by *Natura*, just as the gifts of the soil are in farming. It is also clear that a universal concept of medicine is possible, allowing for both simple and compound uses of *remedia*, and for foreign as well as Roman practices.

We may add some other factors which blur Pliny's rigid divisions. Certain herbal ideas are opposed to rational ideas and discoveries. In *HN* 25. 10, for instance, Pliny admits that, despite the existence of scientific explanations of eclipses, of which, as we have seen, he approves (Ch. II. 3), there still exists among the *vulgus* a belief that they are due to the power of charms and magic herbs. We saw, in Ch. III. 2. A. ii, that the idea of the 'power' of herbs linked herbal medicine inextricably with magic. This was true, not only for Greece, but also for Italy. 'Circe of Italy' features in Pliny's account of herbal medicines (25. 11). The rather dubious reputation of her descendants, the Marsi, the contemporary Italian experts in such knowledge, must have been an embarrassment to Pliny's advocacy of natural remedies (25. 11, cf. 7. 15).

The reluctance of herbalists to communicate the secrets of their power (cf III. 2. A. iv) was against Pliny's principles of the freedom of Nature's gifts to all men in need of them, and was, moreover, against the principles of scientific medicine generally. Scribonius Largus complains that, if physicians refrain from using drugs in order to deny others their use, they are even more culpable, because they burn with envy (*Comp.*, pref. 3). We might refer to the competition between doctors, or speculate on the suppression of knowledge by more eminent physicians to preserve their reputation. Yet in Pliny the only reference to this grudging attitude with regard to what ought to be free is found in connection with his herbal medicine. *Invidia* is, on rare occasions, to be found in *Natura* with regard to the provision of remedies for man. Certain animals are grudging of the products which may be obtained from them (see. Ch. IV. 3. D). But it is the *invidia* of man towards man with regard to the benefits of life which earns particular condemnation from Pliny (18. 1–5, Ch. V. 1. B. ii). Refusal to communicate herbal remedies in effect also puts those responsible on the same level as animals generally, who find their own cures by haphazard chance discovery. They are denying the blend of *ratio* and *usus* which is peculiar to man and which provides by transmission a unique unity, not only between man and man, but also between one generation and another (*HN* 27. 7–8; see Ch. II. 2). As we saw earlier (Ch. VI. 2. B), Pliny's implications of impermanence and lack of continuity in the theory of the rival medical sects suggest that Greek medicine, too, did not meet these standards of true knowledge. In fact, the Greek medical literature gave it an inestimable advantage in this respect over the unorganized, illiterate herbal tradition as described by Pliny himself (25. 16ff.).

It does not do to forget that common assumptions existed behind all ancient ideas of healing. Sympathy and antipathy are accepted as principles by doctors generally, and also by Pliny, in particular with regard to medicine (24. 1, 37. 59). They were considered to be rational ideas with empirical proof, but were equally at home in magic.[49] One of Pliny's own passages on

[49] Cf. Edelstein, *Papers* (n. 5), 232 ff. Pliny is not short of examples (20. 1): water puts out fire, and the sun absorbs water, while the moon gives it birth. Both these heavenly bodies suffer eclipse by each other. The magnet draws iron. These

sympathy and antipathy provides an ironic proof of the futility of trying to compartmentalize Greek medicine, herbal medicine, and magic. When discussing the medical remedies to be obtained from gemstones, he comments that, in gems, *naturae maiestas* is gathered into the narrowest of limits (37. 1). At the same time, he is anxious to criticize the extravagant claims of the Magi, since 'in very many of their statements about gems they have gone beyond a pleasant substitute for medicine into the realms of the supernatural (*ad prodigia transgressi*)' (37. 54). Yet, in *HN* 37. 59–60, he affirms his own belief in sympathy and antipathy, adducing in proof of the principle the notion that diamond can be broken up by goat's blood. Furthermore, as we have already seen (Ch. II. 2), this was a phenomenon so bizarre that Pliny actually discounts all rational explanation in favour of *numinum voluntas*. When Pliny himself enters into the 'realm of the supernatural' to explain natural phenomena, there is all the more reason to be sceptical of his attempts to create watertight divisions between herbal medicine and magic. Such attempts only show more clearly how closely all ideas of healing and magic were interconnected in the mind of the average Roman. In *HN* 26. 18, for example, he complains that Asclepiades was helped above all by 'Magian deceits', which proved strong enough to destroy all confidence in herbal medicine. He then pours scorn on the exaggerated claims made by the Magi for certain herbs. Yet these are no less believable than the properties he is inclined to accept in *HN* 25. 14–15 as evidence of *vis naturae* (see Ch. III. 2. A. iv).

Did the popularity of Pliny's books on medicine bear any relation to the general decline in a rational approach to medicine from the second century onwards? The passage in *HN* 37. 59 ff. is the only occasion on which Pliny appears to adopt a supernatural explanation in place of his more rational inclinations. It is not to be taken as typical of his thought. Yet, together with the magical nature of many of the herbal remedies he records, it may have appealed to users of his work in a less rational era. Echoes of Plinian ideas and even phrases are to be found in the fragments of the *Physica* of Theodorus Priscianus, physician to the emperor Gratian. They are numerous enough to suggest some familiarity

effects he would have seen, even if the same cannot be said of the effect of goat's blood on diamond (see Ch. II. 2; VI. 2. c).

with the *HN*, at least through an intermediate source.[50] Edelstein considered that these writings typified a later Christian reluctance to pursue medical inquiry into the mysteries of the natural world.[51] Priscianus found Plinian phrases with which to express the power and mystery of Nature: *naturae vis*, *natura omniparens*, and *natura qua operetur grande secretum* (*Physica*, 2 Rose) are closely related to the language of *HN* 2. 101; 7. 7; 18. 1, 5; 37. 205. However, ideas which in Pliny had been expressed ambiguously or even with open derision are now set down with uncritical confidence: opinions of Democritus and Pythagoras, whom Pliny had been inclined to view with doubt, as being fanciful (2. 84) or inclined to magic (30. 10), are quoted. Pythagoras is credited with a theory of sympathy and antipathy, but it is used to explain without comment a custom Pliny explicitly condemns in *HN* 2. 16, Roman deification of *mala* as in their dedication of a temple to Fever (*Phys.* 3 Rose). To Democritus is attributed a magical remedy for epilepsy with a typically magical justification; he stressed the power (*potentia*) of the remedy and forbade investigation of the theory (*ratio*). Here, then, is a possible example of a later author applying his own preconceptions to Pliny's text, and producing a distortion of his ideas. In the case of the users of his passages on animals, we saw in Ch. IV that the misrepresentation was quite serious. In the case of the power of herbal remedies, however, it must be admitted that there are more grounds for the later interpretation. The interest of both Pythagoras and Democritus in the properties of plants is specifically recorded by Pliny in the passage where, despite magical connotations, he is at his most enthusiastic about herbal 'power' (25. 13). The tenor of Priscianus' reference to Democritus is reminiscent of *HN* 37. 60, exceptional though that may be. Moreover, descriptions of magical plants are given in *HN* 24. 156–67 on the authority of these two authors. While Pliny implicitly disclaims responsibility for the stories by writing in indirect speech and referring to his longer discussion of the

[50] Many parallels with the *HN* are cited by Rose in the Teubner edition of 1897, as are references to the later *Medicina Plinii*, which was probably taken from the *HN* c.300–50: see A. Önnerfors, *In Medicinam Plinii Studia Philologica* (Lund, 1963).

[51] On Priscianus, see Edelstein, 'Greek Medicine in Its Relation to Religion and Magic', *Papers* (n. 5), 205–40, esp. 235.

Magi in later books of the *HN* (25.167), a hasty reader or excerptor would not notice this.

D. HERBAL MEDICINE: PLINY'S PERSONAL INTEREST

His advocacy of herbal remedies thus shows Pliny at his least rational and nearer than he would wish to admit to ideas of which he does not approve. Was it necessary to push this advocacy to such extremes simply to fulfil an ideal of man living in harmony with nature or to express general humanitarian principles? Even the divinity of Nature herself was adequately attested by all the other marvels of the natural world recorded in the thirty-seven volumes of the *HN*. Dislike of Greek intellectual pretensions does not seem adequately to explain the declaration of such an artificial division within the art of healing either. It is therefore worth examining another factor not so far considered: Pliny's personal interest in the study of plants. From what little evidence we can glean from his writing, botanical study seems to have been the one area of natural history which Pliny had the time and opportunity to pursue, however superficially, in person. He tells us in *HN* 25.9 that he has been able to examine all but a few plants in the *hortulus* of Antonius Castor, 'who was the greatest authority of our era in that field'. The inclusion of details of Castor's great age and good health implies that he was a living testimony to the success of cures from his plants. He may have made a great impression on Pliny, who rarely suggests that he has met his sources in person.

It is easy to denigrate the depth and seriousness of Pliny's own botanical studies.[52] He probably did conduct his investigations

[52] Analyses of Pliny's botany generally conclude that he is less reliable than his chief source, Theophrastus. However, they also admit that he can be excused his errors to a certain degree. J. Stannard ('Pliny and Roman Botany', *Isis*, 56 (1965), 420–5) distinguishes between inaccuracies and botanical errors. He and J. André ('Pline l'Ancien botaniste', *REL* 33 (1955), 297–318) point to inaccuracies in Theophrastus and Dioscurides, and to cases where Pliny has described a plant more accurately than Theophrastus (e.g. 21.23 on the Madonna lily, according to Stannard). His incompetence is most marked when dealing with exotic plants, which he would not have seen, and other errors are the result of methods of working; thus reading by another leads to mishearing (André). Stannard does lay stress on the evidence in the *HN* for personal observation, and the value of the information thus obtained. Caecilius Largus showed him nettle-trees (17.5); mint turns yellow in autumn (19.159); he is puzzled why it should be thought that a

with the attitude of the avid collector or amateur bird-watcher. He sends to the 'experts' for rare plants: 'I have met a herbalist doctor who said that the plant Moly is also found in Italy and could be brought in a few days from Campania' (25. 27). What is remarkable is that, despite all his other commitments and all the other topics covered in the *HN*, he was able to show any practical interest at all. Idealistic and patriotic feelings would have encouraged him to choose this field for particular attention, but the convenience of Castor's *hortulus*, with its comprehensive collection of 'all but a few' plants, must have been at least as important in this respect. It would hardly be surprising if Pliny accentuated his tendency towards herbal medicine for the reasons already discussed in this chapter, if it was an area in which he could claim a certain personal *auctoritas*. This point will be returned to at the end of this section.

The importance Pliny would thus attach to personal contact with the basic raw materials of the medical *ars* could well have had good practical as well as idealistic reasons. To know the ingredients of a remedy as it were theoretically, rather than by personal observation, might have real dangers, misunderstandings of the type Pliny describes in *HN* 29. 25: he has known doctors to add a poison, red lead, known in Latin not only as *minium* but also as *cinnabaris* (33. 116), to a remedy, because they confused the name with that of Italian cinnabar. Pliny's attacks on 'theory' seem more understandable when seen in the light of this anecdote. It is not very much use for a doctor to be able to make complex mixtures like the Mithridatic antidote to precise measurements, if his second-hand knowledge leads to such potentially fatal errors.[53] Pliny certainly recognizes the

triplex is difficult to grow in Italy (20. 220). See also relevant articles by P. Rovesti and A. Scarpa in *Plinio e la natura*, 87–92, 79–85; A. G. Morton in *Science in the Early Roman Empire* (n. 13), 86–97, and R. Meiggs, *Trees and Timber in the Ancient Mediterranean World* (Oxford, 1982), 24–5, 420–2.

[53] A difference in emphasis between Hippocratic pharmacologists and the root-cutters is again of interest here. Lloyd (n. 46) 134 ff. suggests that the former were rationalistic in explanation but careless about plant identification: there is not the same care with identification as is to be found in Theophrastus, a main source for Pliny. A related problem which is not explicitly stated by Pliny, but which may have occupied him in the context of compound drugs and imported remedies, is that of the adulteration of such substances by intermediaries, such as local traders. Galen was aware of this on his travels looking for remedies. On his visit to Lemnos, he was concerned to establish that no blood was added to the medicinal earth, *terra sigillata*: see O. Temkin (n. 47), 14 and n.

usefulness of herbal handbooks in spreading knowledge of their appearance and uses; yet it must also be remembered that his comment on his visit to Castor's garden comes as a consequence of remarks on the difficulty of describing plants to another person. Identification from books, whether they include pictures or rely on detailed verbal descriptions (25. 8), may not always be correct.

This problem, moreover, may have been of contemporary significance. Pliny considers that the turning away from the field to the lecture-room was a fairly recent medical aberration. He comments on a successful blending of herbal knowledge with Greek theory by Greek doctors in earlier times. The rot set in only after Herophilus, who himself used herbs, 'although he was the founder of an over-subtle sect' (26. 10–11). Pliny may well be concerned at the situation in contemporary Rome. A vast population, combined with comparatively limited areas of wasteland, must have made personal contact of either patient or doctor with the raw ingredients in their natural habitat more and more impracticable.[54] When he says that old customs have perished because of the greatness—*magnitudo*—of the Roman People (24. 5), Pliny may well be thinking of magnitude of numbers as well as of power. It was easier to obtain ready-prepared drugs, and the picking of the plants themselves may well have become largely confined to the humble root-gatherers, as Pliny suggests in *HN* 25. 16.

A near-contemporary anecdote brings in another practical drawback of the drug-trade. A familiar Plinian attitude also comes in to play: imported drugs, like imported luxuries, are liable to be subject to price-fluctuations and be sold at rates which do not reflect their true value. Shipping-costs, the terms on which a particular merchant has bought them, or the domination of the market by one dealer, may all send prices very high. Under Nero, the entire Seplasia, the well-known druggists' district of Capua, prosecuted the dealer Demetrius, according to Pliny (33. 164), for such malpractice.

Finally, we should return to the idea of Pliny as an authority on herbal medicine. His own interest, and his intention to encourage an interest in others, was not merely an eccentricity on his part.

[54] See V. Nutton (n. 13), 39.

Medicine was regarded as an important element of the Greek *paideia*,[55] a topic on which the informed layman, as well as the expert, could and should have an opinion. As such, it was an obvious subject for Pliny to treat in his encyclopaedic work (see Intro., pp. 11–13). Moreover, his interest may be seen as part of a particular tendency to encourage lay interest in medicine which began in the first century BC and continued in the first century of our era.[56] A celebrated piece of didactic poetry also belongs to this era: Nero's physician Andromachus described his antidote in verses addressed to the emperor (Galen xiv. 32, 211, 270 Kühn). Another physician-poet, Damocrates (Galen xiv. 32 Kühn), a contemporary of Pliny's, is mentioned in the *HN* for his verses publicizing the uses of the plant *hiberis* (25. 87, cf. 24. 43). Thus Pliny's polemical tone may be at least in part due to a wish to establish a distinctive position in a matter of topical and general interest. His remarks about lecture-rooms and the Roman People losing their practical knowledge of traditional herbal remedies may be prompted by the existence of lectures for laymen as well as by the more obvious context of the training of *medici*. Athenaeus of Attaleia, founder of the Pneumatic school, who practised at Rome, possibly in Claudius' principate, commended such lay interest.[57] He specifically mentions attendance at medical lectures. Pliny wished the layman's study of medicine to concentrate on practical knowledge of herbals, not on the Greek ideas, and in this he set a practical example. However, we could go further, and suggest that, as he wished to vanquish his rival, Columella, in the field of farming, so, to a certain extent, he saw himself as supplanting all previous Roman writers on herbal medicine, even the revered Cato. In *HN* 25. 4 ff., he says that Cato only touched briefly on the subject. The book written by one C. Valgius and dedicated to Augustus was left unfinished. Finally, P. Lenaeus' work was merely a translation of Mithridates'. It now becomes clearer why those Romans who embraced Greek medical doctrines are to all intents and purposes ignored

[55] See W. Jaeger, *Paideia* (Oxford, 1944), iii. 11–21.

[56] F. Kudlien (n. 4), who also mentions the lost work of Rufus of Ephesus called *For Laymen*, cf. Dioscurides' and Galen's *Euporistica*, works on remedies easily found by laymen.

[57] Athenaeus' treatise *On Healthy Living* from Oribasius, *Corp. Med. Graec.* vi. 2. 2., pp. 138–41. See Kudlien (n. 4), 20.

in the *HN*, even when, like Celsus, they wrote in Latin. The anti-Greek stance is all the more explicable if we believe that Pliny wishes to establish himself as a, or even the, Roman authority on medicine, in opposition to the Greeks and their ideas.

E. *HUMANA SALUS*: HEALING HERBS

We saw that there are some practical points in Pliny's advocacy of personal knowledge of herbal remedies. It can be seen that *usus* is particularly important in medicine, where mistaken recognition is literally a matter of life and death. The extremity of the situations with which medicine has to deal must also be borne in mind when attempting to understand Pliny's extreme stance. *Humana salus* was, as we saw in the first part of this chapter, central to his ideology. In general, ancient literature suggests that the doctor was the subject of extreme reactions, depending on whether the patient lived or died. Seneca's gratitude to doctors, the accusations of murder on tombstones (29. 11), and the humour derived from the figure of the incompetent and unscrupulous doctor in satire and comedy, are all aspects of a real fear of illness.[58] Elsewhere, Pliny is eager to advance rational explanations to rid people of irrational fears (Ch. II. 3). In the matter of medicine, however, the learned Hellenistic science, with its rational explanations, in actual fact could probably claim no more cures than less costly and pretentious herbal remedies. Greek medicine generally and Hellenistic medicine in particular is often said to have suffered above all from this rationality, from a tendency to prefer *a priori* hypotheses to empirical evidence. To this extent, Pliny's preferences can be justified. There was also too little systematization of knowledge about the causes of diseases and the concept of infection.[59] All this would add to Pliny's cynicism with regard to the misguided ingenuity of the Greeks generally.

[58] Also found in epigram, e.g. *AP* 11. 114, 116, 125.

[59] O. Temkin, *The Double Face of Janus and Other Essays in the History of Medicine* (Baltimore, 1977), 456–71, 'An historical Analysis of the Concept of Infection', shows that the ancients knew of the concept of contagion, a prime example being Thucydides' description of the plague at Athens in 430 BC (2. 47–54). However, they did not elaborate it systematically. See also J. Solomon, 'Thucydides and the Recognition of Contagion', *Maia*, NS 37 (1985), 121–3.

Pliny appreciates and even shares the common fear of illness. The only passage mentioned so far in which he doubts the pleasantness of *vita* for man is that in *HN* 7. 1 ff., which has more than a touch of the rhetorical commonplace (Ch. II. 3). In *HN* 25. 23, however, the pessimism with regard to a disease-ridden human existence is altogether more spontaneous. 'To reflect on all the different kinds of diseases makes one pity the lot of man; besides chances and changes and every kind of strange happening which every hour brings, there are thousands of diseases to be feared by each single mortal.' Then comes a remark which sums up the commom experience. 'To decide which are the most serious of them might be considered almost as an act of stupidity, since each man considers the disease from which he is presently suffering to be the most horrible . . .'. The multiplicity of different diseases is an added burden of fear for a human race already threatened every hour by chance calamities. The coupling of disease with the unpredictable in life also acts as a reminder that fear of disease was often unmitigated by knowledge of their causes. What wrath of the gods, he asks in *HN* 26. 9, has added new diseases to the three hundred or so already known, as though man had not enough to fear as it was? Pliny, however, does not despair, although he refrains from stressing the advantages of *ratio*, so badly misused by the pernicious Greeks. Instead, he advocates a more basic understanding of *Natura*. He implicitly recommends a greater systematization of herbal medicine in *HN* 25, when he praises Mithridates (§§ 5 ff.), regrets the problems of passing on accurate information in the form of illustrated handbooks (§ 8), and deplores the secrecy and illiteracy of rustic knowledge (§§ 16 ff.). This in fact fits his usual blend of *ratio* and *usus*, even if the overt emphasis is on the latter only.

HN 25. 23 is also another manifestation of the Plinian 'hypochondria' postulated in sect. 1. B, the less obvious, negative, aspect of Pliny's interest in *vita*. Despite the positive attempt to counteract the fear of disease by providing practical information on remedies, there remain a number of passages in which death is seen as the only cure for physical illness. While Pliny is anxious about the risk of accidental death from taking dangerous drugs (sect. 1. C), he accepts the use of poison by those who wish to end their sufferings by suicide (25. 24). He records that opium was used in this way by the father of P. Licinius Caecina, a man of praetorian rank, 'when an intolerable disease had made life hateful to him' (20. 199), and by several others. His castigation of those who use foul remedies from human bodies (Ch. III. 2. A. iii; above, sects. 1. C, 2. B) includes the following reminder: 'wherefore let every man consider that first among the remedies of his soul is this: that of all the blessings which Nature has bestowed on man, none is greater than a timely death and the best thing about it is that each man has the power to produce it for himself' (28. 9). The Utopian Hyperboreans (Ch. II. 4) achieve the ideal in this respect: it was their custom to commit suicide when satiated with good living at a ripe old age. Thus, they avoided mortal sickness (4. 89). Even God cannot commit suicide, 'the greatest gift he has given to man in a life of such tribulation' (2. 27, see Ch. I. 1). It is noticeable that Pliny seems to be interested in suicide solely as a means of escaping physical ills and makes no mention of it as a way of resolving political or moral compromises, a feature which had particular relevance to the so-called 'Stoic Opposition' of his own times. Perhaps anything which might be construed as a reference to the fate of Thrasea Paetus, who was condemned by Nero and took his own life, would have been tactless. His son-in-law, Helvidius Priscus, had been causing trouble for Vespasian and was in fact executed by the emperor at the very time when Pliny was writing the *HN* (*c*.75). This lack of reference to suicide for the sake of principle is certainly interesting. Perhaps it is a purely coincidental omission. It may, on the other hand, be a manifestation of Pliny's practical, even materialistic view of 'life'.[60] Finally, he also offers a grim reminder that there was little ancient medicine could do to alleviate the general debilitation of old age: in *HN* 7. 168 ff., he expresses indignation that this should be designated a period of 'life' at all.

[60] Contrast Seneca's more high-minded approach in *Ep*. 58. 36, where he says he will not commit suicide 'propter dolorem'. See A. Bodson, *La Morale Sociale des derniers stoïciens, Sénèque, Epictète et Marc Aurèle* (Paris, 1967), 101. M. T. Griffin, 'Philosophy, Cato and Roman Suicide', *G&R*, NS 33 (1986), 64–77, 192–202, esp. 200, suggests Pliny gives us an unusual glimpse of the more commonplace Roman suicide.

Some Conclusions

In the introduction to this chapter, it was suggested that Pliny's attitude to the arts of healing would bring together the main themes of his thought as a fitting conclusion to this study. It would, therefore, be inappropriate to end on a note of uncharacteristic pessimism. Pliny never tires of emphasizing the divine power of *Natura*, but it is the power of a life-giving deity, whose *remedia* far outweigh her poisons (18. 5). Her provision of the mainstays of human survival, food and medicine, is her greatest work (20. 1, cf. 19. 189; 27. 1–2). Even her most inhospitable areas yield these gifts. 'Not even the woods and the wilder side of Nature lack medicines, there being no place where that sacred mother of all things did not distribute remedies for mankind, so that the very desert was made a source of medicine' (24. 1). The farm, with its association of a harmonious relationship between man and nature, is even more significant in providing the most familiar and easily obtainable remedies (sect. 1. A). Yet it is Pliny's portrait of the garden which in many ways epitomizes the harmonious and optimistic partnership between man and nature. Here could be cultivated all plants that were of benefit to *salus humana* and all harmful products eliminated. It is appropriate that the garden should yield a compound medicine that is approved by Pliny, ironic that it is Greek. A verse inscribed on stone in the temple of Asclepius at Cos describes how to make lozenges from nine different herbs and the best wine. They make a comprehensive antidote to all the poisonous animals which are a reproach to *Natura*'s normal benevolence towards man (20. 264).[61]

[61] The only exception is the asp. Pliny attributes the recipe to King Antiochus.

BIBLIOGRAPHY

THE bibliography contains only such works on Pliny as are relevant to the themes of this book, together with those general works which have a bearing on particular topics. For a more extensive survey of Plinian studies, the reader is referred to the bibliographies by Sallmann and Serbat as listed below, though it should be noted that these are by no means exhaustive.

ABT, A., *Die Apologie des Apuleius von Madaura und die antike Zauberei* (Giessen, 1908).

ANDRÉ, J. M., 'Pline l'Ancien botaniste', *REL* 33 (1955), 297–318.

—— 'Erreurs de traduction chez Pline l'Ancien', *REL* 37 (1959), 203–45.

—— 'Nature et culture chez Pline', in id. (ed.), *Recherches sur les* artes *à Rome* (Publ. Univ. de Dijon, 58; Paris, 1978), 7 ff.

—— 'L'epidémiologie de Pline', *Helm.* 37 (1986), 45–52.

ARMSTRONG, A., 'The Methods of the Greek Physiognomists', *G&R*, NS 5 (1958), 52–6.

ARNOLD, E. V., *Roman Stoicism* (Cambridge, 1911).

ASTIN, A. E., *Cato the Censor* (Oxford, 1978).

AUGUET, R., *Cruelty and Civilisation: The Roman Games* (trans., London, 1972).

BALDRY, H. C., *The Unity of Mankind in Greek Thought* (Cambridge, 1965).

BALSDON, J. P. V. D., 'Sulla Felix', *JRS* 41 (1951), 1–10.

—— *Romans and Aliens* (London, 1979).

BARB, A. A., 'The Survival of Magic Arts', in A. Momigliano (ed.), *Paganism and Christianity in the Fourth Century* (Oxford, 1963).

BEARD, M., and CRAWFORD, M., *Rome in the Late Republic* (London, 1985).

BEAUJEU, J., 'La cosmologie de Pline l'Ancien dans ses rapports avec l'histoire des idées', *REL* 26 (1948), 40–1.

BESSONE, L., 'Sulla morte di Plinio il Vecchio', *Rivista di studi classici*, 17 (1969), 166–79.

—— 'Plinio e i due Rufi: Virginio e Cluvio', *Helm.* 38 (1987), 135–44.

BIANCHI, E., 'Teratologia e geografia', *Acme*, 34 (1981), 227–49.

Bidez, J., and Cumont, F., *Les Mages héllenisés: Zoroastre, Ostanes et Hystaspe d'après la tradition grecque*, 2 vols. (Paris, 1938).
Binder, G., 'Auguste d'après les informations de la *N.H.*', *Helm.* 38 (1987), 145–56.
Bloch, R., *Recherches sur les religions de l'antiquité classique* (Paris, 1980).
Bodson A., *La Morale sociale des derniers stoïciens, Sénèque, Épictète, et Marc Aurèle* (Bibl. de la faculté de philos. et lettres de l'Univ. de Liège, 176; Paris, 1967).
Bodson, L., 'Aspects of Pliny's Zoology', in *Science in the Early Roman Empire*, 98–110.
—— 'La zoologie romaine d'après la *N.H.* de Pline', *Helm.* 37 (1986), 107–16.
Bouché-Leclercq, A., *L'astrologie grecque* (Paris, 1899).
Bowersock, G. W., *Greek Sophists in the Roman Empire* (Oxford, 1969).
Brenk, F. E., *In Mist Apparelled; Religious Themes in Plutarch's Moralia and Lives* (Leiden, 1977).
Bruère, R. T., 'Pliny and Virgil', *C. Phil.* 51 (1956), 228–46.
Brunt, P. A., review of K. D. White, *Roman Farming*; *JRS* 62 (1972), 153–8.
—— *Italian Manpower 225 B.C.–A.D. 14* (Oxford, 1971, repr. 1987).
Burns, M. A., 'Pliny's Ideal Roman', *CJ* 69 (1964), 233–8.
Cagnolaro, L., 'I mammiferi marini di Plinio', in *Plinio e la natura*, 27–38.
Capelle, W., 'Die Schrift von der Welt', *Neue Jahrbücher*, 15 (1905), 632.
Capitani, U., 'Celso, Scribonio Largo, Plinio il Vecchio e il loro attegiamento nei contrasti della medicina popolare', *Maia*, NS 24 (1972), 120–40.
Capponi, F., *Le fonti del X libro della Naturalis Historia di Plinio* (Univ. di Genova, fac. di lett., Publ., ist. di filol. class. e med.; Genoa, 1985).
—— 'Cultura scientifico-naturalistica di Plinio', *Helm.* 37 (1986), 131–46.
Caprotti, E., 'Animali fantastici: Fantasie zoologiche e la loro realtà in Plinio', in *Plinio e la Natura*, 39–61.
Carnoy, A. J. 'Magic (Iranian)', *Encyclopaedia of Religion and Ethics*, ed. J. Hastings (Edinburgh, 1908–26), viii, 293–6.
Caspar, J. W., *Roman Religion as Seen in Pliny's Natural History* (Chicago, 1934).
Casson, L., *The Ancient Mariners* (London, 1959).
—— *Travel in the Ancient World* (London, 1974).
—— *The Periplus Maris Erythraei* (Princeton, 1989).

CHADWICK, H. (tr. and comm.), *Origen: Contra Celsum* (Cambridge, 1965).

CHAMPEAUX, J., *Fortuna: Recherches sur le culte de la Fortune à Rome et dans le monde romain des origines à la mort de César*, i (Col. éc. Fr. de Rome, 64; Paris, 1982).

CHARLESWORTH, M. P., *Trade Routes and Commerce of the Roman Empire* (Cambridge, 1926).

CHEVALLIER, R., 'Le bois, l'arbre et la forêt chez Pline', *Helm.* 37 (1986), 147–72.

CHIBNALL, M., 'Pliny's *Natural History* and the Middle Ages', in T. A. Dorey (ed.), *Empire and Aftermath* (London, 1975).

CITRONI MARCHETTI, S., '*Iuvare mortalem*: L'ideale programmatico della *N.H.* di Plinio nei rapporti con il moralismo stoico-diatribico', *A&R* 27 124–48.

COHN-HAFT, L., *The Public Physicians of Ancient Greece* (Smith Coll. Stud. Hist.; Northampton, Mass., 1956).

COLSON, F. H., introduction to *De Aeternitate Mundi* in the works of Philo of Alexandria, Loeb edn., ix (Cambridge, Mass., 1941).

COSSARINI, A., 'Latifundia perdidere Italiam', in *Plinio il Vecchio*, 253 ff.

CRAMER, F. H., *Astrology in Roman Law and Politics* (Mem. Amer. Phil. Soc. 37 Philadelphia, 1954).

CRAWFORD, M., 'Economia imperiale e commercio estero', in *Tecnologia*, 207–17.

CUMONT, F., *The Oriental Religions in Roman Paganism*, trans. (New York, 1956).

DANNEMAN, F., *Plinius und seine Naturgeschichte* (Jena, 1921).

DELLA CASA, A., 'Plinio grammatico', in *Plinio il Vecchio*, 109–15.

DESANGES, J., 'Les sources de Pline dans sa description de la Trogodytique et de l'Éthiopie (*H.N.* 6. 163–97)', *Helm.* 37 (1986), 279–92.

DICKISON, S., 'Abortion in Antiquity', *Arethusa*, 6 (1973), 159–66.

DIERAUER, U., *Tier und Mensch im Denken der Antike* (Amsterdam, 1977).

DIHLE, A., 'Plinius und die geographische Wissenschaft in der römischen Kaiserzeit', in *Tecnologia*, 121–37.

DODDS, E. R., *Pagan and Christian in an Age of Anxiety* (Cambridge, 1965).

—— *The Ancient Concept of Progress and Other Essays on Greek Literature and Belief* (Oxford, 1973).

DOVER, K., 'The Freedom of the Intellectual in Greek Society', *Talanta*, 7 (1976), 24–54.

DRINKWATER, J. E., *Roman Gaul* (London, 1985).

DUBUISSON, M., 'Les Opici: Osques occidentaux ou barbares?',

Latomus, 42 (1983), 522–45.

Ducos, M., 'Existe-t-il une philosophie du droit chez Pline?', *Helm.* 37 (1986), 201–18.

Dulaey, M., 'La notice de Pline sur les Esséniens (5. 73)', *Helm.* 38 (1987), 283–93.

Dumont, J. C., 'Quelques aspects de l'esclavage et de l'économie agraire chez Pline', *Helm.* 37 (1986), 293–306.

Dumont, J.-P., 'L'idée de Dieu chez Pline (*HN* 2. 1–5, 1–27)', *Helm.* 37 (1986), 219–37.

Edelstein, E. J., *The Hippocratic Oath* (Baltimore, 1943).

—— and Edelstein, L., *Asclepius*, 2 vols. (Baltimore, 1945).

Edelstein, L., *The Meaning of Stoicism* (Martin Class. Lects. 21; Cambridge, Mass., 1966).

—— *The Idea of Progress in Classical Antiquity* (Baltimore, 1967).

—— *Selected Papers of L. Edelstein*, ed. O. Temkin (Baltimore, 1967).

Erkell, H. D., *Augustus*, *Felicitas*, *Fortuna*; *Lateinische Wortstudien* (Göteborg, 1952).

Ernout, A., 'La magie chez Pline l'Ancien', in M. Renard and R. Schilling (eds.), *Hommages à J. Bayet* (Collection Latomus, 70; Brussels, 1969), 190–5.

Evans, E. C., 'The Study of Physiognomy in the Second Century A.D.', *TAPA* 72 (1941), 96–108.

—— 'Galen the Physician as Physiognomist', *TAPA* 76 (1945), 287–98.

—— 'A Stoic Aspect of Senecan Drama', *TAPA* 81 (1950), 169–84.

Farrington, B., *Greek Science*, rev. edn. (London, 1963).

Ferguson, J., *Religions of the Roman Empire* (London, 1970).

Festugière, A. J., *La Révélation d'Hermès Trismégiste*^{3}, i (Paris, 1950).

—— *Personal Religion among the Greeks* (Berkeley, Calif., 1954).

Fischer, K. D., 'Quelques réflexions sur la structure et deux nouveaux témoins de la *Physica Plinii*', *Helm.* 37 (1986), 53–66.

Forbes, C. A., 'The Education and Training of Slaves in Antiquity', *TAPA* 86 (1955), 321–60.

Förster, R., *Scriptores Physiognomici Graeci et Latini*, 2 vols. (Leipzig, 1893).

Fortenbaugh, W. W., 'Aristotle: Animals, Emotion and Moral Virtue', *Arethusa*, 4 (1971), 137–63.

Frank, T. (ed.), *An Economic Survey of Ancient Rome* (Baltimore, 1934–40).

Frederiksen, M., 'Plinio il Vecchio e l'agricoltura in età imperiale romana: Gli aspetti tecnici ed economici', in *Tecnologia* (Como, 1980), 81–97.

French, R., 'Pliny and Renaissance Medicine', in *Science in the Early Roman Empire*, 252–81.

FÜHRMAN, M., *Das systematische Lehrbuch* (Göttingen, 1960).

GAGÉ, J., *Matronalia* (Coll. Lat. 60; Brussels, 1963).

GAROSI, R., 'Indagine sulla formazione del concetto di magia nella cultura romana', in P. Xella (ed.), *Magia: Studi di storia delle religioni in memoria di Raffaela Garosi* (Rome, 1976), 13–93.

GEYTENBEEK, A. C. VAN, *Musonius Rufus and Greek Diatribe* (Assen, 1963).

GIANNINI, A., *Paradoxographorum Graecorum Reliquiae* (Milan, 1965).

GIGON, O., 'Plinius und der Zerfall der antiken Naturwissenschaft', *Arctos* 4 (1966), 23–45.

—— 'Pline', in *Plinio il Vecchio*, 41–52.

GOTTHELF, A. (ed.), *Aristotle on Nature and Living Things: Philosophical and Historical Studies* (Bristol, 1985).

GOUREVITCH, D., 'Asclépiade de Bithynie dans Pline: problèmes de chronologie', *Helm.* 37 (1986), 67–82.

GREEN, P., 'Prolegomena to the Study of Magic and Superstition in the *Natural History* of Pliny the Elder, with Special Reference to Book xxx and its Sources' (diss., Cambridge, 1952).

GRIFFIN, M. T., *Nero: The End of a Dynasty* (London, 1984).

—— 'Philosophy, Cato and Roman Suicide', *G&R*, NS 33 (1986), 64–77, 192–202.

—— *Seneca: A Philosopher in Politics* (London, 1976).

GRIMAL, P., 'Encyclopédies antiques', *Cah. d'hist. mond.* 9 (1965), 459–82.

—— *Les Jardins romains à la fin de la République et aux deux premiers siècles de l'Empire*[2] (Paris, 1969).

—— 'Pline et les philosophes', *Helm.* 37 (1986), 239–49.

GRISÉ, Y., 'L'illustre mort de Pline le Naturaliste', *REL* 58 (1980), 338–43.

GRMEK, M. D., 'Les circonstances de la mort de Pline: commentaire medical d'une lettre destinée aux historiens', *Helm.* 37 (1986), 25–43.

GRÜNINGER, G. *Untersuchungen zur Persönlichkeit des älteren Plinius: Die Bedeutung wissenschaftlicher Arbeit in seinem Denken* (diss. Freiburg, 1976).

GUDGER, E. W., 'Pliny's *Historia Naturalis*: The Most Popular Natural History Ever Published', *Isis*, 6 (1924), 269–81.

GUITTARD, C., 'Pline et la classification des prières dans la religion romaine (*N.H.* 28. 10–21)', *Helm.* 38 (1987), 157–70.

GUTHRIE, W. K. C., *In the Beginning; Some Greek Views on the Origins of Life and the Early State of Man* (London, 1957).

—— *A History of Greek Philosophy*, iii (Cambridge, 1969).

HADOT, I., *Seneca und die griechische-römische Tradition der Seelenleitung* (Berlin, 1969).

HAYS, H. R., *Birds, Beasts and Man; A Humanist History of Zoology*

(London, 1973).

Haywood, R. M., 'The Strange Death of Pliny the Elder', *Class. Week.* 46 (1952), 1–3.

Healy, J., *Mining and Metallurgy in the Greco-Roman World* (London, 1978).

—— 'Pliny on Mineralogy and Metals', *Science in the Early Roman Empire*, 111–46.

Hild, J. A., 'Fortuna-Tyche', in Daremberg–Saglio ii. 2. 1264–77 (Paris, 1896).

Hine, H. M. (ed. and comm.), *Seneca, Natural Questions*, Book 2 (New York, 1981).

Hopkins, K., 'A Textual Emendation in a Fragment of Musonius Rufus: A Note on Contraception', *CQ*, ns 15 (1965), 72–4.

—— 'Contraception in the Roman Empire', *Comparative Studies in Society and History*, 8 (1965–6), 124–51.

—— *Death and Renewal* (Cambridge, 1983).

Howe, N. P., 'In Defense of the Encyclopedic Mode: On Pliny's Preface to the *H.N.*', *Latomus* 44 (1985), 561–76.

Isager, G., *Pliny on Art and Society: The Elder Pliny's Chapters on the History of Art* (London, 1991).

Jacob, C., and Lestringant, F. (eds.), *Arts et légendes d'espaces, figures du voyage et rhétoriques du monde* (Paris, 1981).

Jaeger, W., *Early Christianity and Greek Paideia* (Harvard, 1962).

—— *Paideia*, 3 vols. (Oxford, 1944).

Jal, P., 'Pline et la historiographie latine', *Helm.* 38 (1987), 171–86.

Janson, T., *Latin Prose Prefaces* (Stockholm, 1964).

Jashemski, W. J., *The Gardens of Pompeii*, i (New York, 1979).

Jennison, G., *Animals for Show and Pleasure in Ancient Rome* (Manchester, 1937).

Jex-Blake, E., and Sellars, E., *The Elder Pliny's Chapters on the History of Art* (Oxford, 1896).

Jones, W. H. S., 'Ancient Documents and Contemporary Life, with Special Reference to the Hippocratic Corpus, Celsus and Pliny', in E. A. Underwood (ed.), *Science, Medicine and History*; *Essays in Honour of Charles Singer* (London, 1953), i. 100–10.

Kajanto, I., 'Fortuna', *ANRW* II. 17. 1 (Berlin and New York, 1981), 502–58.

Keller, O., *Rerum Naturalium Scriptores Graeci Minores* (Leipzig, 1877).

—— *Die antike Tierwelt* (Leipzig, 1909; repr. Hildesheim, 1963).

Kolendo, J., 'L'agricoltura nell'Italia romana in Plinio il Vecchio', in *Plinio il Vecchio*, 109–11.

König, R., and Winckler, G. (eds.), *Plinius der Ältere: Leben und*

Werk eines antiken Naturforschers (Munich, 1979) [*non vidi*].

KÖVES-ZULAUF, Th., *Reden und Schweigen; römische Religion bei Plinius Maior* (Munich, 1972).

—— 'Plinius der Ältere und die römische Religion', *ANRW* II. 16. 1 (Berlin and New York, 1978), 187–288.

KROLL, W., 'Plinius und die Chaldäer', *Hermes*, 65 (1930), 1–13.

—— *Die Cosmologie des Plinius* (Breslau, 1930).

—— 'Plinius 5', *RE*, xxi. 1 (1951), 271–439.

KUDLIEN, F., 'Medical Education in Classical Antiquity', in C. D. O'Malley (ed.), *The History of Medical Education* (Berkeley, Calif., 1970), 1–37.

LABHARDT, A., 'Quelques témoignages d'auteurs latins sur la personnalité et l'œuvre de Pline l'Ancien', in *Mélanges offerts à Max Niedermann* (Neuchâtel, 1944), 105–15.

—— 'Curiositas', *Mus. Helv.* 17 (1960), 206–26.

LANA, I., 'Scienza e politica in età imperiale romana', in *Tecnologia*, 21–43.

LATTE, K., *Römische Religionsgeschichte* (Munich 1960), 176–83.

LAUSDEI, C., 'La *NH* di Plinio: Problemi esegetici e linguistici', *Helm.* 38 (1987), 255–71.

LAWSON, J., 'The Roman Garden', *G&R* 19 (1950), 97–105.

LE BŒUFFLE, A., 'Pline et l'astrologie', *Helm.* 37 (1986), 173–84.

LEE, D., 'Science, Philosophy and Technology in the Greco-Roman World', *G&R* 42 (1973), 65–76, 180–93.

LEE, H. D. P., 'Place Names and the Date of Aristotle's Biological Works', *PCPS* 179 (1946–7), 7–9.

LEITNER, H., *Zoologische Terminologie beim älteren Plinius* (Hildesheim, 1972).

LESKY, A., *Thalatta, der Weg der Griechen zum Meer* (Vienna, 1947).

LEWIS, P. R., 'Roman Gold-Mining in North-West Spain', *JRS* 60 (1970), 169–85.

LEY, W., *Dawn of Zoology* (New York, 1968).

LIEBESCHUETZ, W., *Continuity and Change in Roman Religion* (Oxford, 1979).

LLOYD, G. E. R., *Greek Science after Aristotle* (London, 1973).

—— *Magic, Reason and Experience: Studies in the Origins and Development of Greek Science* (Cambridge, 1979).

—— *Science, Folklore and Ideology* (Cambridge, 1983).

—— (trans.), *Hippocratic Writings* (London, 1983).

LOCHER, A., 'The Structure of the *Natural History*', in *Science in the Early Roman Empire*, 20–9.

LOISEL, G., *Histoire des ménageries de l'Antiquité à nos jours* (Paris, 1912).

Long, A. A., *Problems in Stoicism* (London, 1971).
Lovejoy, A. O., and Boas, G., *Primitivism and Related Ideas in Antiquity* (Baltimore, 1935).
MacDougall, E. B., and Jashemski, W. J. (eds.), *Ancient Roman Gardens* (Dumbarton Oaks Colloquium on the History of Landscape, 7; Washington, DC, 1981).
—— —— (eds.), *Ancient Roman Villa Gardens* (Dumbarton Oaks, Washington, DC, 1987).
McElderry, R. K., 'Vespasian's Reconstruction of Spain', *JRS* 8 (1918), 94–101.
Martin, R., 'La mort étrange de Pline l'Ancien, ou l'art de la déformation historique chez Pline le Jeune', *VL* 73 (1979), 13–21.
—— *Recherches sur les agronomes latins et leurs conceptions économiques et sociales* (Paris, 1971).
Mazzini, I., 'Presence de Pline dans les herbiers de l'antiquité et du haut Moyen-Age', *Helm.* 37 (1986), 83–94.
Meiggs, R., *Trees and Timber in the Ancient Mediterranean World* (Oxford, 1982).
Meijer, F., *A History of Seafaring in the Classical World* (London, 1986).
Mendieta, E. Amand de, 'The Official Attitude of Basil of Caesarea as a Christian Bishop towards Greek Philosophy and Science', *Studies in Church History*, 13 (1976), 25–49.
Mette, H. J., 'Enkyklios Paideia', *Gymnasium*, 67 (1960), 300–7.
Michel, A., 'L'esthétique de Pline l'Ancien', *Helm.* 38 (1987), 55–67.
Millar, F., *The Emperor in the Roman World* (London, 1977).
Miller, J. I., *The Spice Trade of the Roman Empire* (Oxford, 1969).
Moellering, H. Arnim, *Plutarch on Superstition* (Boston, 1963).
Morin, J., *Le Dessin des animaux en Grèce d'après les vases peints* (Paris, 1911).
Morton, A. G., 'Pliny on Plants; his Place in the History of Botany', in *Science in the Early Roman Empire*, 86–97.
Münzer, F., *Beiträge zur Quellenkritik der Naturgeschichte des Plinius* (Berlin, 1897).
Nauert, C. G., Jnr., 'Humanists, Scientists and Pliny: Changing Approaches to a Classical Author', *Am. Hist. Rev.* 84 (1979), 71–85.
Nilsson, M. P., 'The New Concept of the Universe in Later Greek Paganism', *Eranos*, 44 (1944–6), 20–7.
—— *Geschichte der griechischen Religion*, ii^3 (Munich, 1967).
Nock, A. D., *Essays on Religion and the Ancient World*, ed. Z. Stewart, 2 vols. (Oxford, 1972).
North, J. A., 'Conservatism and Change in Roman Religion', *PBSR* 54 (1976), 1–12.
—— 'Novelty and Choice in Roman Religion', *JRS* 70 (1980), 186–91.

NUTTON, V., 'The Perils of Patriotism: Pliny and Roman Medicine', in *Science in the Early Roman Empire*, 30–58.

ODER, E., and HOPPE, C. (eds.), *Corpus Hippiatricorum Graecorum* (Teubner, 1924–7, repr. 1971).

ÖNNERFORS, A., *Pliniana: In Plinii Maioris Naturalis Historia—Studia Grammatica, Semantica, Critica* (Uppsala, 1956).

—— *In Medicinam Plinii studia philologica* (Lund, 1963).

OROZ RETA, J., 'Présence de Pline dans les *Étymologies* de Saint Isidore de Séville', *Helm.* 38 (1987), 295–306.

PAASSEN, C. VAN, *The Classical Tradition of Geography* (Groningen, 1957).

PACK, R. A., 'Artemidorus and the Physiognomists', *TAPA* 72 (1941), 321–34.

PANI, M., review of *Tecnologia: Quaderni di Storia*, 18 (1983), 361–70.

PASTUCCI, G., 'La lettera prefatoria di Plinio alla *N.H.*', in *Plinio il Vecchio*, 171–97.

PATCH, H. R., *The Tradition of the Goddess Fortuna in Roman Literature and in the Transitional Period* (Smith Coll. Stud. in Mod. Languages, 3/3; Northampton, Mass., 1922).

PEASE, A. S., 'Caeli enarrant', *Harv. Theol. Rev.* 34 (1941), 163–200.

PEDERSEN, O., 'Some Astronomical Topics in Pliny', *Science in the Early Roman Empire*, 162–96.

PETROCHILIS, N. K., *Roman Attitudes to the Greeks* (Athens, 1974).

POHLENZ, M., *Die Stoa*[4] (Göttingen, 1970).

PURCELL, N., 'Wine and Wealth in Ancient Italy', *JRS* 75 (1985), 1–19.

—— 'Town in Country and Country in Town', in E. MacDougall (ed.), *Ancient Roman Villa Gardens* (Dumbarton Oaks, Washington, DC, 1987).

RAMIN, J., *Le Périple d'Hannon* (BAR, suppl. ser., 3; Oxford, 1976).

—— *Mythologie et géographie* (Paris, 1979).

RAWSON, E., *Intellectual Life in the Late Roman Republic* (London, 1985).

RENARD, M., review of P. Grimal, *Les Jardins Romains*, *AC* 14 (1945), 226–8.

RENEHAN, R., 'The Greek Anthropocentric View of Man', *HSCP* 85 (1981), 239–59.

RIST, J. M., *Stoic Philosophy* (Cambridge, 1969).

—— 'Basil's "Neoplatonism": Its Background and Nature', in P. Fedwick (ed.), *Basil of Caesarea, Christian, Humanist, Ascetic*, i (Toronto, 1981), 137–220.

ROBBINS, F. E., *The Hexaemeral Literature* (Chicago, 1912).

ROSTOVTZEFF, M., *The Social and Economic History of the Roman Empire*[2] (Oxford, 1957).

Rovesti, P., 'Attualità erboristica di Plinio il Vecchio', in *Plinio e la natura*, 87–92.

Rutland, L. W., '"Fortuna sola invocatur": Pliny's statement', *Class. Bull.* 56 (1979), 28–31.

Sabbah, G., 'Présence de la *N.H.* chez les auteurs de l'antiquité tardive: L'exemple d'Ammien Marcellin, de Symmaque et d'Ausone', *Helm.* 38 (1987), 203–21.

Saint-Denis, E. de, *Le rôle de la mer dans la poésie latine* (thesis, Lyon, 1935).

Sallmann, K., *Die Geographie des älteren Plinius in ihrem Verhältnis zu Varro: Versuch einer Quellenanalyse* (Berlin and New York, 1971).

—— 'Plinius der Ältere 1938–1970', *Lustrum*, 18 (1975), 5–299; 345–52.

—— '"Quo verius tradere posteris possis" (Plin. epist. 6, 16)', *WJA*, NF 5 (1979), 209–18.

—— 'La responsabilité de l'homme face à la nature', *Helm.* 37 (1986), 251–66.

—— 'Reserved for Eternal Punishment: The Elder Pliny's View of Free Germania (*HN* 16. 1–6)', *AJP* 108 (1987), 108–28.

—— 'Der Traum des Historikers; Zu den *Bella Germaniae* des Plinius und zur julisch-claudischen Geschichtsschreibung', *ANRW* II. 32. 1 (Berlin and New York, 1984), 578–601.

Sambursky, S., *The Physical World of Late Antiquity* (London, 1962).

—— *The Physical World of the Greeks* (London, repr. 1963).

—— *The Physics of the Stoics* (London, 1959).

Sandbach, F. M., *The Stoics* (London, 1975).

Scarborough, J., *Roman Medicine* (London, 1969).

Scarpa, A., 'Le piante medicinali nella *HN* di Plinio il Vecchio', in *Plinio e la natura*, 79–85.

Schilling, R., 'La place de Pline l'Ancien dans la littérature technique', *R.Ph.* 52 (1978), 272–83.

Schmidt, J., 'Physiognomik', *RE* xx. 1 (Stuttgart, 1941), 1064–74.

Schoff, W. H., *The Periplus of the Erythraean Sea* (London, 1912).

Schrödinger, E., *Nature and the Greeks* (Oxford, 1954).

Serbat, G., 'La référence comme indice de distance dans l'énoncé de Pline l'Ancien', *R.Ph.* 47 (1973), 38–49.

—— 'Il y a Grecs et Grecs! Quel sens donner au prétendu antihéllénisme de Pline?', *Helm.* 38 (1987), 272–82.

—— 'Pline l'Ancien: État présent des études sur sa vie, son œuvre et son influence', *ANRW* II. 32. 4 (Berlin and New York, 1986), 2069–200.

Shaw, B. D., 'The Elder Pliny's African Geography', *Historia*, 30 (1981), 424–71.

SHERWIN-WHITE, A. N., *Racial Prejudice in Imperial Rome* (Cambridge, 1967).

—— 'Pliny, the Man and His Letters', *G&R*, NS 16 (1969), 76–90.

SINGER, C., *Greek Biology and Greek Medicine* (Oxford, 1922).

——'The Herbal in Antiquity and its Transmission to Later Ages', *JHS* 47 (1927), 1–52.

—— and UNDERWOOD, E. A. (eds.), *A Short History of Medicine*² i (Oxford, 1962).

SKYDSGAARD, J. E., *Varro the Scholar* (Copenhagen, 1968).

SMITH, K. F., 'Magic (Greek and Roman)', *Encyclopaedia of Religion and Ethics*, ed. J. Hastings (Edinburgh, 1908–26), viii. 268–89.

STAHL, W. H., *Martianus Capella and the Seven Liberal Arts*, 2 vols. (New York, 1971).

STANNARD, J., 'The Plant called Moly', *Osiris*, 14 (1962), 254–307.

—— 'Pliny and Roman Botany', *Isis*, 56 (1965), 420–5.

—— 'Medicinal Plants and Folk Remedies in Pliny, *H.N.*' *HPLS* 4 (1982), 3–23.

—— 'Herbal Medicine and Herbal Magic in Pliny's Time', *Helm.* 37 (1986), 95–106.

STEINER, G., 'The Scepticism of the Elder Pliny', *Class. Week.* 48 (1955), 137–43.

SUTHERLAND, C. H. V., *The Romans in Spain 217 B.C.–A.D. 117* (London, 1939).

SYME, R., 'Carrière et amis consulaires de Pline', *Helm.* 38 (1987), 223–31.

TEMKIN, O., *Galenism: Rise and Fall of a Medical Philosophy* (Ithaca, NY, and London, 1973).

—— *The Double Face of Janus and Other Essays in the History of Medicine* (Baltimore, 1977).

TERIAN, E., *Philonis Alexandrini de Animalibus*, Armenian text with introduction, translation, and commentary (Berkeley, Calif., 1981).

THÉRASSE, J., 'Croyances et crédulités des Romains, d'après Pline l'Ancien et les écrivains latins', in J. B. Caron, M. Fortin, and G. Maloney (eds.), Mélanges Lebel (Quebec, 1980), 283–319.

THIEL, J. H., *Studies on the History of Roman Sea-Power in Republican Times* (Amsterdam, 1946).

THOMAS, K., *Man and the Natural World: Changing Attitudes in England 1500–1800* (London, 1984).

—— *Religion and the Decline of Magic* (London, 1971).

THOMSON, J. O., *History of Ancient Geography* (London, 1948).

TOYNBEE, J. M. C., *Animals in Roman Life and Art* (London, 1937).

—— review of P. Grimal, *Les Jardins Romains*; *JRS* 36 (1946), 210–13.

TUPET, A.-M., *La Magie dans la poésie latine* (Paris, 1976).

Vegetti, M., 'Zoologia e antropologia in Plinio', in *Plinio il Vecchio* (Como, 1982), 117–31.

Vogel, C. J. de, *Greek Philosophy*, iii (Leiden, 1959).

Wallace-Hadrill, A., 'Pliny the Elder and Man's Unnatural History', *G&R*, ns 37 (1990), 80–96.

Wallace-Hadrill, D. S., *The Greek Patristic View of Nature* (Manchester, 1968).

Wardman, A., *Rome's Debt to Greece* (London, 1976).

—— *Religion and Statecraft among the Romans* (London, 1982).

Warmington, E. H., and Cary, M., *The Ancient Explorers*² (London, 1963).

Weinstock, S., review of K. Latte, *Römische Religiongeschichte*; *JRS* 51 (1961), 206–15.

—— *Divus Julius* (Oxford, 1971).

Westlake, H. D., review of A. Lesky, *Thalatta*; *CR* 63 (1949), 118–19.

White, K. D., 'Latifundia', *BICS* 14 (1967), 62–79.

—— *Roman Farming* (London, 1970).

Wightman, E. M., *Gallia Belgica* (London, 1983).

Williams, G., *Change and Decline: Roman Literature in the Early Empire* (Berkeley, Calif., 1978).

Wilson, N. G., *St. Basil on Greek Literature* (London, 1975).

Winspear, A. D., *Lucretius and Scientific Thought* (Montreal, 1963).

Wistrand, E., *Felicitas Imperatoria* (Göteborg, 1987).

Wolters, X. M. F. G., *Notes on Antique Folklore on the Basis of Pliny H.N. xxviii. 22–29* (Amsterdam, 1935).

Wolverton, R. E., 'The Encomium of Cicero in Pliny the Elder' (Classical Medieval and Renaissance studies in honour of B. L. Ullman, i; Rome, 1964), 159–64.

Xella, P. (ed.), *Magia: Studi di storia delle religioni in memoria di Raffaela Garosi* (Rome, 1976).

Zehnacker, H., 'Pline l'Ancien, lecteur d'Ovide et de Sénèque', in H. Zehnacker and G. Hentz (eds.), *Hommages à R. Schilling* (Paris, 1983), 437–46.

Zirkle, C., 'The Death of C. Plinius Secundus 23–79 a.d.', *Isis*, 58 (1967), 553–9.

INDEX